# MASSIVELY SMALL

# MASSIVELY SMALL

STELIOS PAMFILIS

CreateSpace, a Division of Amazon,
Massively Small

Editor: Matthew Coyne.
EMCH



Printed in the United States of America

First Printing, 2016

ISBN-13: 9781530372409
ISBN-10: 1530372402
Library of Congress Control Number: 2016904173
CreateSpace Independent Publishing Platform
North Charleston, South Carolina

*Dedicated to my children*

# TABLE OF CONTENTS

# *Chapter 1*

## THE OLD ORDER

*In the beginning, God created the heaven and earth. And the earth was without form and void; and darkness was upon the face of the deep.*

—*Genesis 1:1–2*

It is human nature to seek answers to ultimate questions. For as long as man has had self-consciousness, he has wondered also about the origins of the universe. Most of our attempts have entered the realm of the metaphysical, which may be beyond our grasp. It has given us comfort to think that the forces of nature and even the fate of time are pushed by God (or the gods). Our ancient theories, including those calling on Zeus, Hades, and Poseidon, fell out of favor thousands of years ago and have since been replaced by other theories.

Some theories are based on faith, some on general observation, and some on mathematical equations. All of these reiterations have their values and pitfalls. Faith-based explanations are beyond the scope of this book. Whatever the driving force may be, we will restrict our thoughts to the measurable mechanisms that have resulted. We will therefore make an attempt to make sense of the world around us with the existing scientific theories and data available. We have made great strides in the last century in attempting to categorize the details of the mechanisms by which the universe operates.

We turn to science to help explain the world in ways that are more accessible to our limited human minds. Science has allowed us to make great strides in our attempts to work out the goings-on around us. As we discover more, it leads to new insight and accessibility to this understanding. Over the millennia, scientific theories were brought forth, improved upon, discounted, replaced, and refined. Typically, these theories start as intuition based on our common, everyday experiences. Right or wrong, a theory that fits with our everyday experience tends to be readily accepted. As it turns out, unfortunately, when it comes to the universe at its smallest and biggest scales, our common intuition tends to be pretty darn awful.

The most ambitious scientific undertakings have sought a single-system explanation for everything. Looking at our environment, with its organisms, forces, and people, it was the inclination of early scientists to qualify and categorize their surroundings. As early as the eighteenth century BC, the cosmos was fully accountable by five basic elements: sea, earth, sky, fire, and wind.[1]

In the Western culture we are more commonly taught about the Greek interpretation. This contribution is often attributed to Plato as he put forth a similar system, characterizing and listing the basic elements for us: fire, air, water, earth.[2]

In the fourth century BC, Aristotle further described the qualities and behaviors of these elements. He added one more. The fifth essential element makes up that which is beyond our tangible world. Possibly it is the stuff stars are made of. This fifth essence, the quintessence, he called "aether."

Similar theories were also developed independently in the Eastern tradition. In the Hindu text *Tattwa Kaumudi*, the same elements are listed. In this formulation, instead of these elements being essential, one is built on the next. The most essential element, akasha (aether), can form the next most complex element, air. Akasha and air can form fire. Akasha, air and fire in combination creates water. The most complex level of these elements, earth, is the mixture of all the preceding elements. These levels correspond to the increasing levels of complexity of our senses: sound, feel, vision, taste, smell.

Ancient traditions up until recent times held these four plus one essential elements. Even now we have the somewhat similar notion of four states of matter: gas, liquid, solid, plasma. In the past century, the idea of the aether has been more or less dismissed. But it may be a worthwhile exercise to ask if there is still room for a quintessential element.

The history of science is also a history of systematization. Plato, Aristotle, and their predecessors all thought they truly understood their surroundings. Of course, we feel today that the positions of the Greeks and those before them do not accurately explain the world exactly as it works. In building these descriptive and theoretical frameworks, these quasi-scientists were building a discourse, a shared language and process, through which they could measure and predict the world around them. For all intents and purposes, their theories were correct insofar as the theories allowed them to accurately predict the behavior of the world around them. If you observe that the Sun rises each morning and sets each evening, yet you believe it is because the Sun orbits the Earth, you are still able to predict its behavior tomorrow and a month from now. If you're a theoretical physicist writing a textbook, your factually incorrect understanding matters in that it will hurt your reputation. But if you're a farmer, and your understanding allows you to rise with the Sun each day, that's probably good enough.

Perhaps the most comprehensive and influential account of the process by which dominant scientific theories replace one another is put forth by Thomas Kuhn in his 1962 masterwork in the history and philosophy of science, *The Structure of Scientific Revolutions*. Kuhn famously describes the now-familiar concept of paradigms as "universally recognized scientific achievements that for a time provide model platforms and solutions for a community of practitioners." A key word in that definition is "practitioner." Scientific paradigms allow people to *practice* science. Scientific paradigms are *practical*.

Closely related to the Kuhn's concept of paradigms is his concept of "normal science." In Kuhn's definition, normal science is "research firmly based upon one or more past scientific achievements, achievements that some particular scientific community acknowledges for a time as supplying the foundation for its further practice."[3] Further, normal science is "predicated on the assumption that the scientific community knows what the world is like. Much of the success of the enterprise derives the community's willingness to defend that assumption."[4] Here it is important to note that, while it is easy to take the existence of paradigms for granted, Kuhn emphasizes that different fields reach the level of consensus required for paradigms at different rates. Getting to the first paradigm in a field is arduous work. But for the intents of our discussion in this book, we can assume that, in physics, this initial level of arduous work has been done, and we can thus accept the existence of paradigms and of normal science in this field.

No matter how entrenched and "normal" it gets, a paradigm will always stand on shaky ground because "an apparently arbitrary element, compounded of personal and historical accident, is always a formative ingredient of the beliefs espoused by a given scientific community at any time." A 1933 profile of Einstein, though it predates Kuhn by thirty years, presages this idea, noting, "It is possible that, if Einstein had been a passionate patriot, a gregarious being, and an orthodox religionist, he might have developed the Einstein theory...[but] it is conceivable that, if Einstein had been a good German, a good Jew, and a good mixer, the relativity theory might not yet have been discovered."[5] The uniqueness of Einstein's personality was as important a factor in the scientific revolutions of the twentieth century as anything else. This arbitrariness, which gives birth to a paradigm, will in turn cause its death and give birth to a new paradigm. One eccentric innovator eventually usurps another.

It is only in the period in between that we are blind to the arbitrariness of the entire theoretical enterprise. Just as all political revolutions must eventually shift their focus to the dull task of governing, all scientific revolutions eventually (and necessarily) slide into the work of normal science—at least until the next Copernicus or Newton or Lavoisier or Einstein comes along.

Kuhn's work reaffirms that it is good to question theories that don't quite seem to make sense. All of the ancient theories that were held so dearly and passed on as fact have been replaced with new theories. We cannot assume that what we are taught as fact today will stand up to tomorrow's scrutiny. There is plenty of room for skepticism. While it may be frustrating to live as a contrarian, always suspecting the opposite to be true, there is a joy in discovering things for oneself. It is good to question the authority of one's teachers and the validity of their lessons. It is fair to wonder if the material one has to regurgitate back on a test, and on which one's grade depends, is even vaguely similar to actual truth. Hundreds of years ago, these taught "truths" were all completely different. Hundreds of years from now, today's "truths" may also be laughable.

It is good exercise to question theories that don't seem to make sense. In contemporary physics, theories of photons, gravitons, motion, uncertainty principle, matter, space, and time all have major inconsistencies. All beg further clarification. It is within this framework that I ask you to share my skepticism and seek alternatives that hopefully make a little more sense.

We look out the window and see trees and hills and valleys. Earth's surface may have some character, but in general, it goes a long way in every direction in a pretty level manner. As a percentage of the Earth's size, the topography is actually quite flat. Mount Everest rises 29,029 feet above sea level. Compare this to the average radius of the Earth, which is 21,000,000 feet. Put in that perspective, Mount Everest represents just over a 0.1 percent change. It's easy to see why with limited scientific resources that it seemed that the Earth was, in general, flat. Even so, it was still quite ancient times when some intellectuals found contrary evidence. Some slowly adopted newer theories. Among those theories, some believed that the world existed as a disk. In response to the experience of watching ships in the distance fall beneath the horizon, our Earth shape theory progressed to more of a dome. We could see different stars from different inclinations from different positions. The Earth casts a round shadow on the moon during a lunar eclipse. This was good evidence of a spherical Earth. The flat Earth should have been easy to let go of, but still it took centuries to do so. Thus it is with many scientific revolutions; evidence slowly accumulates in favor of a major paradigm shift. Even so, no matter how glaringly obvious the accumulated body of evidence looks in retrospect, it often takes a single major breakthrough or "aha" moment to fundamentally change the conversation.

Although the shift from a flat Earth to a sphere may have been a jarring one, at least we felt secure in the fact that the Earth was the center of the universe. So secure that we incorporated this into our religious beliefs. It became heresy to suggest otherwise. Despite our willingness to torture and kill others for thinking and speaking their minds, we eventually came to a point when we had to let go of our geocentric viewpoint for the more logical heliocentric universe. This was acceptable with some minor adjustments to our religious understanding of the placement of God's throne. We now also trust that the Sun is not the center of the universe, either.

It was valuable to go through these various iterations and have them evolve into our present understanding, however imperfect. It was wrong then, and still is, to think of any given understanding as anything more than an evolving theory. Each was an incomplete theory at best. Theory-in-flux. Theory under construction. The theories are useful insofar as they allow us to predict the world around us. They have been and will be replaced with more expansive theories that will allow for broader or more accurate predictions.

We see this same scientific evolutionary process in every field. Medicine is especially prone to changing theories. We scoff at our ancients, for instance, who used leeches to purge the evil humors from the infirmed body. Thankfully medicine has come a long way since then, but many would be surprised to learn that leeches are still in medical use for skin infections, burns, and grafts. The natural anticoagulant hirudin, which leeches excrete in their saliva, is used for medical applications. It is now made by recombinant genetics and acts as an antifibrinogen. It is more potent than the common anticoagulant heparin. Even if the recombinant DNA source makes us feel more clinically secure, it is still that leech saliva molecule infusing into the bloodstream. The practitioners of old were onto something, even if their explanations and observations had nothing to do with our present understanding of genetics.

In Newton's *Opticks*, as in many theories prior, an underlying frame of reference was assumed. This was the luminiferous aether. It was the logical explanation for some of the observed behavior of light and heat. As a side note, it should be pointed out that he later suggested that such a medium was unlikely to exist, as it would tend to retard the movement of the planets. For the next two hundred years or so, the theories of electromagnetism were built in the setting of a presumed aether.

It was only a little over a hundred years ago that the scientific dogma still held strong to the concept of luminiferous aether. It was assumed that objects moved in the universe in reference to this stationary framework. Toward the end of the nineteenth century, as instrumentational precision improved, many of the attributes associated with the aether were disproved. The Michelson-Morley experiment, done in 1887 in Cleveland, Ohio tested for the existence of aether by observing the speed of light in different directions through the presumed stationary aether. Their negative results were a strong suggestion that maybe there is no aether. Other similar experiments then confirmed their findings and again were interpreted as proof against the existence of an aether.

The Michelson-Morley experiment most famously influenced none other than Einstein himself, whose work on special relativity seemed to put the final nail in the aether's coffin. In 1905, the theory of special relativity suggested that an object's motion only makes sense in reference to another object. There is no a priori motion. When two objects are in nonaccelerated motion with respect to each other, each has the rightful claim to being the stationary reference or the object in motion or a little of both. There is no essential positional framework necessary to make this theory work. This was the moment that the aether was abandoned, as the lack of necessity in this new powerful schemata of special relativity does not require it. This alone was felt to be sufficient evidence to disprove its existence.

Yet after several decades of discounting of the aether, we joyously celebrated the Higgs mechanism. The Higgs mechanism involves an invisible field that reaches into every corner of the cosmos. The Higgs ocean, wrapped up in the jargon of modern physics is much more acceptable to us than the simplistic, intuitive idea of the aether that has been around for so many centuries.

The Higgs mechanism is one of many concepts in modern physics that, when considered with an open mind, is reminiscent of Aristotle's aether. They have in fact had many iterations in many civilizations; waves in their aether media, quantum fields, Higgs field, dark energy, cosmologic constant, strings, branes. Though it may seem silly to consider these newer concepts using a millennia-old archaic concept, every present-day scientific theory borrows aspects of this aether in some fashion or another. We feel better, more scientific, more modern, and more sophisticated when we rename it. The term "aether" itself is simply not palatable in modern physics. However, although it may be unacceptable, disavowed, laughable, scorned, it still remains, renamed and reinvented, as an essential element in how we think things work.

As we noted at the start of this chapter, most of what we suspect to be true of the universe is based on logical extrapolation from our everyday experience. These events seem to us to be the normal stuff. Well, as it turns out, our everyday experience here on Earth is actually the weird exception when it comes to how things work at the very smallest essential scale or at the very largest, the astronomical scale. It will be necessary to let go of these notions of what is "normal" in order to see the truth. As we learn in *Caddyshack*, "In one physical model of the universe, the shortest distance between two points is a straight line in the opposite direction."

Reductionism and unification are noble scientific endeavors to help systematize the universe at these smallest and largest scales. Reductionism most certainly dates back before any reliable history would account for, but in the historical record, it has been attributed to Thales from what is now Miletus, Turkey, in about 450 BC. He stated that all things are ultimately made of a single substance. He even said it in a different language, which made it sound even more profound.

A science historian, Gerald Holton, coined the term "Ionian Enchantment" referring to the Ionian Sea as the general region of this school of thought. This particular school of thought lasted centuries and looked specifically for unity in the forces of nature. For Thales, it seemed most likely that the essential building block was water. Later, Anaximander, of the same school of thought but of a different generation, felt that there must be some other essence, an infinity or apeiron that constitutes all of the other essential elements that are familiar to us.

What this early school of thought proposed has been reiterated for centuries and still holds true today. The way nature works, the way the universe is organized, very complex stuff is made by varying the patterns of simpler building blocks. The smaller the units get, the simpler they become until we arrive at the simplest of all building blocks. Every once in a while, we think we have arrived at that smallest possible unit. The atom was for a long time felt to be the smallest possible unit. "Atom" even means "unsplittable." It is indeed amazingly small, but we now all firmly believe in the subatomic particles named protons, neutrons, and electrons. These generic "subatomic" units are the same for all atoms. It is only in the way they are configured that offers the variety of atoms in existence. A hydrogen proton is identical to a uranium proton. At each level the building blocks fall into even more simple possibilities that make up more complex matter. And on it goes to the sub-subatomic particles of the standard model and possibly further on to strings. The closer we look, the more essential levels of building blocks we discover.

So let us get something straight that has been incorrectly written in science books for years. The ancient Greeks felt that the universe was composed of essential, unsplittable units called atoms. Nonancient non-Greeks (Dalton, for instance) later suggested that what we now call atoms are those essential units. It was later determined that these were indeed splittable, and thus the ancient Greeks were wrong. But wait—they never said our atoms were their atoms. The ancient Greeks may have been correct in suggesting an unsplittable essential unit, but we were very premature in assigning this important title to what we presently call atoms. Maybe someday, when all the dust settles, we can come up with a more appropriate name for particles at the elemental level so that we can reassign the title "atom" to something truly unsplittable.

The reductionist impulse described above can also be understood as the impulse to simplify. The impulse to simplify leads us to Occam's razor, which holds that the correct explanation is usually the simplest explanation among all of the alternatives. Occam's razor encourages us to choose the "truth" based on the simplest, most straightforward solution. Though this generally holds true for our everyday experiences, I would suggest that Occam also causes a dilemma. The predictability and repeatability of our everyday experiences lead us to believe that what happens at the human scale also holds up at the atomic scale. We also extrapolate our experiences to the astronomic scale. When we try to reconcile the behavior of the universe at the smallest and grandest scales with our intuition and our daily experiences, the behavior of the universe can seem absurd. But it is more likely that our intuition is absurd. We can call this Occam's dilemma: Whatever seems to make sense in our world, the opposite must be true in the rest of the universe.

Physics obviously seeks to explain a broader range of phenomena than the aforementioned subjects of light, electromagnetism, and the building blocks of matter. Two additional key phenomena, which are equal parts familiar and mystifying, are gravity and magnetism. The fact that these two familiar forces act at a distance from the primary object is not particularly disturbing to us. Though we have attempted to explain both phenomena for our entire history, our present theories really don't fully explain these phenomena. In fact, gravity is so poorly understood that we had to invent a new particle, the graviton.

Although our current theory of gravity may be one of the most obviously suspect in all of physics, it may well be time for us to re-explore our notions across the board: light speed, photons, Higgs bosons. We need to let go of our present idea of what mass means.

Let's return to basics. The way nature works, the universe is organized such that very complex stuff is made up of very simple building blocks. We see this again and again. Each block seems to be able to be broken down into even smaller building blocks. With each level of smaller building block, the structure of the unit becomes simpler and simpler of until you finally reached the simplest unit of that particular system.

Take the human body as an example. People are highly organized complex structures. They are composed of about $3 \times 10^{13}$ cells. While all of these cells are unique, they are made from a couple thousand kinds of molecules, fats, carbohydrates, proteins, and nucleic acids, all programmed by the information in these nucleic acids. Except for a spontaneous mutation here and there, an organism's DNA is all identical. All of the DNA is made of only a few kinds of atoms.

Likewise, all of the molecules of the universe are made up of just 118 different atoms. Every atom is made up of some combination of protons, neutrons, and electrons and maybe whatever glues them together. Protons and neutrons might each be made up some of the subatomic particles that populate the standard-model list. It is felt that this list comprises everything that "is." Some theorize further the existence of vibrating strings, loops or variation thereof are an even smaller, simpler level below the standard-model particles. This might be so but it still might not be the lowest order of simplification. We will explore the smallest of scales.

Conversely, we need to put these concepts in the perspective of the largest scales, the universe as a whole. Most of everything is a great expanse of nothingness. Over the entire universe, galaxies take up only a minuscule percentage of space. This is hard to fathom, given that our galaxy would take 100,000 years to cross at light speed, and it is only an average-sized galaxy—and there are an estimated hundred (or hundreds of) billion other galaxies, some of which dwarf us in size.

If you were to grind up all the stars and planets and black holes into atomic dust and scatter it evenly throughout the universe, you would expect one atom in every five cubic meters of space. You would expect 412 million photons per cubic meter and $10^{-29}$ kg of dark energy (in mass-energy equivalents).

In this mathematically diffuse arrangement, there are areas of concentration. Mostly through randomness, areas of concentrated matter happened. As a seed, these areas of concentrated mass exerted their attractive gravitational pull to maintain some sort of integrity.

Comprised of the earliest atoms, approximately three-fourths hydrogen and one-fourth helium, these areas collected to form massive pillars. These clouds of simple atoms are on such a large scale as to dwarfs stars and even galaxies in size. As these collections of atoms swirl and dance, gravity pulls some closer together, applying so much pressure as to create massive balls. With increasing pressure comes increasing temperature and kinetic energy. It is from these pillars that stars and galaxies are born.

Both the subatomic and astrological scales are unfathomable. People are made up of $7 \times 10^{26}$ atoms. Even so you only make up $8.4 \times 10^{-22}$ percent of the total mass of the Earth. Our planet is a mere speck of dust in our galaxy, accounting for $10^{-16}$ percent of the total mass of the galaxy. The Milky Way, being a galaxy of medium size, contains around 300 billion stars. There are estimates of 100 billion galaxies. All the stars, in all these hard to imagine numerous galaxies still only hold 4.9 percent of the total mass of the universe. The other 95.1 percent is felt to exist as 26.8 percent dark matter and 68.3 percent dark energy (Planck Mission Team, Astronomy and astrophysics, 571; A1: arXiv:1303.5062). That alone should make us feel a little less special and not so much the center of the universe.

But even though we can rattle off these concepts, we still know relatively little of the mechanisms that govern the whole enterprise. Like the ancient farmer whose geocentric view of the universe is wrong but lets him predict sunrise and sunset, our current theories in physics are likely limited and ripe for a revolution that allows for a more expansive and more accurate accounting of the universe. Here it is helpful to return to Kuhn's idea of normal science and its tendency to simultaneously flesh out the paradigm of the day but resist the formation of new paradigms:

> Closely examined, [the enterprise of normal science] seems an attempt to force nature into the preformed and relatively inflexible box that the paradigm supplies. No part of the aim of normal science is to call forth new sets of phenomena; indeed those that will not fit into the box are often not seen at all. Nor do scientists normally aim to invent new theories, and they are often intolerant of those invented by others. Instead, normal-scientific research is directed to the articulation of those phenomena and theories that the paradigm already supplies.[6]

The last major paradigm shifts came about a century ago with Einstein's theories of special and general relativity redefining the way we see the depth of the cosmos and with the work in quantum mechanics by Planck and others redefining our understanding of the universe at the smallest levels. Though these two theoretical frameworks are ultimately considered to be fundamentally contradictory of one another (though each has been repeatedly experimentally verified on its own), much of the work since has been to force nature into the "preformed and relatively inflexible box[es]" of relativity or quantum mechanics. Much of this is valuable, worthwhile work that has clarified and expounded upon Einstein's ideas and pushed quantum mechanics forward in very real ways. But there is only so much that can be forced into a box before it is once more worthwhile to look for new sets of phenomena, those that don't fit into the box quite so easily.

Because of the very nature of normal science, change does not come easy. Every revolution in science has been met with reluctance at best, claims of heresy at worst. Kuhn notes that the characteristics of the process of scientific discovery include "the previous awareness of anomaly, the gradual and simultaneous emergence of both observational and conceptual recognition, and the consequent change of paradigm categories and procedures often accompanied by resistance."[7] I understand that resistance is inevitable. But with this book, my goal is to identify and articulate the anomalies in physics as I understand them and, by presenting new ideas in a novel (yet maybe ancient) framework, begin the process of shifting the paradigm once again.

# *Chapter 2*

## REDUCTIO AD ABSURDUM

When Albert Einstein was finally awarded the Nobel Prize in 1921, he remarked in his Nobel lecture that, even after the revolutions brought forth by his theories of special and general relativity, "We seek a mathematically unified theory in which gravitational field and electromagnetic field are interpreted only as different components of the same manifestations of some uniform field." Today, nearly a century later, we have not seen any major revolutions since Einstein and still lack the simple, unified theory he sought, which John Updike, writing in *The New Yorker* referred to as "an encompassing theory of elegant simplicity." While Einstein was certainly not blind to the revolutionary importance of his special and general theories of relativity, he was not satisfied. And further, he was particularly skeptical of the increasing reach of quantum theory, commenting in a private correspondence that "the more success quantum theory enjoys, the sillier it looks."[8] Despite these concerns voiced by Einstein himself, newer theories and extensions of quantum theory have, if anything, grown even more successful and perhaps look sillier than ever. To put it most charitably, there are a host of ideas in present-day physics that do not make sense and that specifically contradict other steadfast pillars of scientific dogma.

Not only has the physics landscape since Einstein mutated into something overly complex and increasingly abstract and abstruse, the path that got us to Einstein was certainly not a simple one, either. For over two centuries preceding Einstein, scientists grappled over whether light was a wave or a particle and whether the universe consisted of or was enmeshed in, as the ancients believed, some sort of all-pervasive medium: an aether. Einstein's work seemed to resolve both debates. First, his work on the photoelectric effect, for which he *was* awarded the Nobel Prize and for which he gave the aforementioned lecture, Einstein led us to the wave-particle duality theory of light, a compromise where light is considered to exhibit behaviors suggestive of both particles and waves. Second, in his work on special relativity, Einstein found that the mathematics of the theory did not require the existence of the aether, and thus he rejected the aether all together. And here we are nearly one hundred years later: we still have the wave-particle duality theory, we still have no aether, and we still have no "encompassing theory of elegant simplicity."

Since Einstein, there have been no true revolutions to fully reconcile and unify theories. To see where we may have gone wrong, and, dare we say it, where Einstein himself may have come up short, it is a worthwhile exercise to return to the idea of the aether. The aether was not an essential component of the paradigm ushered in by Einstein. Because Einstein's theories did not require existence of a quintessence, the new paradigm abandoned it. As Kuhn writes, as a new paradigm arrives, the old ideas tend to disappear because "the new paradigm implies a new and more rigid definition of the field."[9] In this new and more rigid definition of physics, there was no room for aether.

But Einstein's theories do not necessarily *rule out* the existence of the aether. The aether was rejected because it wasn't needed. Einstein, as he should have, chose the simplest theory and did not include unnecessary additives. Here it is important to remember that, while paradigms allow for systematization and some forward progress of science, they can also be restrictive through their necessarily limited views. If a paradigm were to allow the consideration of all facts, it wouldn't be a paradigm at all; it'd be one of the preparadigm "schools" described by Kuhn. However, subsequent ideas that have arisen in physics—string theory, the Higgs-boson particle, the standard model, and even Einstein's theories themselves—may be found to be consistent with the existence of something like the aether, and on this account, it may be worth prodding the soundness of the paradigm, or at least hypothetically broadening its definition to reaccount for the possible existence of the aether. What is needed is a fresh look that takes into account both the wisdom of the ancients and the considerable contributions of modern physicists. Perhaps the "uniform field"

Einstein sought has been with us all this time, and we just need to look at it anew.

The wave theory of light was originally attributed to Christian Huygens in 1672, when he demonstrated the double refraction (birefringence) of calcite crystals, though Huygens had many contemporaries and predecessors who made considerable contributions to this theory as well. His experiments led him to conclude that the only logical way to explain his findings was that light traveled as a wave through a medium, much as sound travels in a wavelike manner through the medium of the air. Light, Huygens writes, "spreads, as Sound does, by spherical surfaces and waves: for I call them waves from their resemblance to those which are seen to be formed in water when a stone is thrown into it, and which present a successive spreading as circles, though these arise from another cause, and are only in a flat surface."[10]

Nearly two hundred years after Huygens made his discoveries about the wavelike behavior of light, James Clerk Maxwell, in his famous equations, laid the mathematical foundation for similar wavelike properties of electric and magnetic phenomena. Maxwell first had his work published when he was fourteen years old, and he would continue to publish on fields as wide-ranging as geometry, color, Saturn's rings, and, of course, electromagnetism, throughout his unfortunately short (forty-eight-year) life. Maxwell worked on electricity both early and late in his life, though by his last decade, despite not fully grasping the implications of all of his work, Maxwell began to sense the far-reaching importance of his work to *all* of physics.

Summarizing his equations, Maxwell conjectured, italicizing for emphasis, that "we can scarcely avoid the inference that *light consists in the transverse undulations of the same medium which is the cause of electric and magnetic phenomena.*"[11] A key word in that inference is "medium." Like Huygens's theory, Maxwell's theory required the existence of a medium, the aether, to work. But if Huygens's medium was a luminiferous aether, Maxwell's was, through its hospitality to light, electric, and magnetic phenomena, a more general aether.

Maxwell's work, though profound in its implications, was lacking in experimental proof and did not fully tie together the findings on electricity and magnetism into a fully unified theory of electromagnetism. Shortly after the publication of Maxwell's equations, however, Heinrich Hertz experimentally proved the validity of Maxwell's equations. Hertz's initial work proved the velocity of electromagnetic waves through air and his later work proved the optical properties (reflection, polarization, etc.) of electromagnetic waves, thus comprising a two-pronged confirmation of Maxwell's theories and mathematical predictions. Maxwell and Hertz, who never met in person, seemed to have established the theoretical and experimental validity of wave theory for good.

These contributions were revolutionary indeed. Early in *The Structure of Scientific Revolutions*, Kuhn writes (almost as an aside), "For the smaller professional group affected by them, Maxwell's equations were as revolutionary as Einstein's, and they were resisted accordingly."[12] Maxwell's equations and Hertz's experiments, most scientific historians would agree, were key foundational concepts for the development of physics for the next century and beyond. In retrospect, the importance of their work is undeniable. But, as Kuhn also notes, the resistance to Maxwell's equations from his contemporaries was fervent, owing largely in part to the centuries-long debate over the wave versus particle theories of light.

While Maxwell and Huygens were strong proponents of the wavelike behavior of light and electromagnetic phenomena, the opposing school led by the towering Sir Isaac Newton (and predating Maxwell by over a century), argued that light was instead composed of particles, or "corpuscles" (which, not coincidentally, seem familiar to our current understanding of photons). Newton's theory held that when a body emits light, it actually emits a stream of tiny particles that then behave mechanically, like any other objects. For example, the corpuscles reflecting off of a smooth mirror are analogous to how a rubber ball would bounce off of a flat surface. One of Newton's most outspoken critics was his contemporary Robert Hooke, a staunch wave-theory advocate. Indeed, the rivalry between the two was so strong, and Newton's fear of Hooke so great, that Newton waited until Hooke died before he published his famous *Opticks*. Partly due to his stature (and perhaps because Hooke was no longer around to vociferously object), Newton's corpuscular theory dominated for over a century. But though Newton's influence still dominated at the turn of the nineteenth century, by 1803 Thomas Young's famous experiments began shifting the argument back toward wave theory.

Young concerned himself with sound waves initially, but it was his strong feeling that in order to account for refraction and color, light must also travel as a wave. This intuition was ultimately confirmed in Young's famous double-slit experiment of 1803. Young arranged an experiment where a single light source was projected through a screen with two openings onto another screen. The result was a series of light and dark bands consistent with an interference patterns. Young's very clear description of the mechanism for this phenomenon was that as the light left the source, it did so in a wave pattern. As this wave reached and passed through the holes in the screen, it created two new points of origin, resulting in two distinct wavelets. These two new waves now hit the back screen. In areas where the two waves struck in harmony, the screen was light, and where these "undulations" were dyssynchronous, there would be a dark band. Young was further able to accurately estimate the wavelength of various colors based on the distance changes in the patterns for different colors with respect to the slit origin.

Leonard Euler, one of the most important mathematicians in history and for whom the natural number *e* is in honor, was another proponent of the wave theory of light and strongly opposed Newton's corpuscular theory. Euler proposed his wave theory, which echoed Huygens, in the 1740s. Given his immense credibility, wave theory regained its superiority and reigned for nearly two centuries.

This brings us back to the era of Maxwell and Hertz. In 1887, Heinrich Hertz bounced energy off a metallic surface. He found that it took a certain threshold of energy to emit the photoelectric effect (the phenomenon where metals appear to emit electrons when being shone upon by light). The interpretation was that this represented light knocking electrons from the metal surface, which generated a current. It was quite confusing, however, that an increase in light intensity (brightness) did not result in more energetic electrons, but instead resulted in more electrons being emitted. Decades later, Albert Einstein proposed discrete energy packets of light (photons) as the explanation for Hertz's findings. Thus, even while Hertz was instrumental in bolstering the wave theory of electromagnetic phenomena, some of his experiments on light further complicated the picture.

As the experimental and mathematical scrutiny of wave theory grew more intense and precise near the turn of the twentieth century, wave theory developed problems of its own. And so new theories, pulling in concepts from both the wave and particle schools, formed. With the strong evidence for both electromagnetic waves and photon particles, a blended duality theory was proposed. Prior to the dual particle-wave theory, the wave vs. particle battle had waged for centuries. Early proponents of wave theory were overpowered by Newton's corpuscles, which reigned for decades until being overthrown by Euler's reformulation of Huygens's wave theory. The Euler/Huygens wave theory then reigned for centuries. Euler was overthrown by Einstein in the early twentieth century, and Einstein's wave-particle duality stills rules today.

These centuries of arguments yielded insights that ultimately led to Einstein's additions to the conversation, which were novel, groundbreaking, and obviously changed physics forever. But since Einstein, the conversation has only grown more muddled. And, barring any sensible intervention, there are no signs of increased clarity in sight. Occam's razor tells us that a wave-particle duality theory does not satisfy the simplicity test. But if Occam's razor tells us that the simplest of all alternatives is likely to be the correct one, what are the alternatives to the duality theory? Perhaps taking a look back at the ancient concepts could provide us with this simpler alternative we so desperately need.

In 1600, the English physicist and intellectual iconoclast William Gilbert coined the word "*electricus*." The word came from the Greek "*electron*," which meant "amber," as a reference to the static electricity created from rubbing amber rods together. In Gilbert's wake, many scientists made brilliant contributions to the development of electric and magnetic theory in the seventeenth, eighteenth, and early nineteenth centuries. Among them were such giants as Robison, Cavendish, Volta, and Priestley. Beginning around the turn of the nineteenth century, a flurry of activity saw Charles-Augustin de Coulomb publish his law defining the relationship of magnetism and distance, Michael Faraday develop the idea of fields to explain how forces act a distance from source material, and James Clerk Maxwell devise his equations that define the behavior of both electric and magnetic phenomena and, furthermore, predict wave propagation of both electric and magnetic occurrences.

Resulting from these efforts was an established body of work that explained the wave behavior of electromagnetic occurrence. Electromagnetic waves could be explained in a predictable, calculable manner. Electromagnetic waves are like any other waves we experience in our lives.

The behavior of waves, electromagnetic and otherwise, is very predictable. Waves are nature's way of transmitting some natural phenomenon from one place to another. All waves perform this function and do so in a predictable, calculable

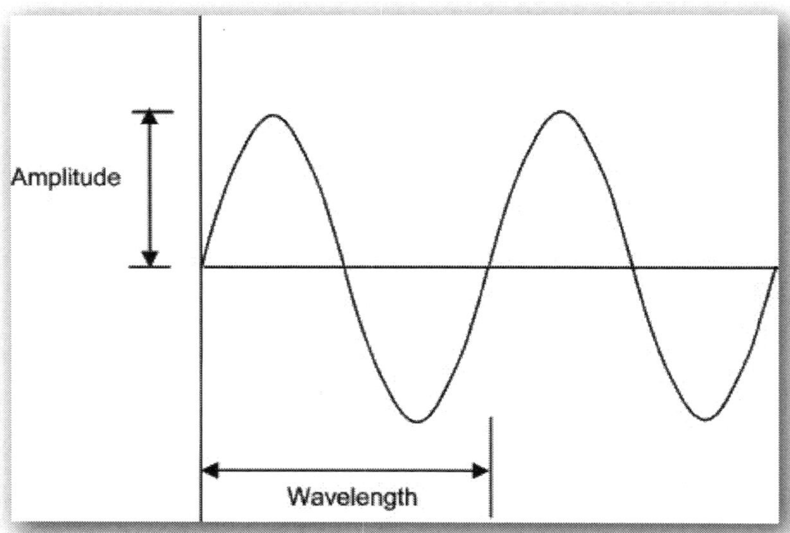


*Figure 1: A wave and its components.*

manner. All waves behave similarly, even though some waves strike our senses and our intuition in differing ways.

In the simplest terms, waves can be described by three measurable components: wavelength, frequency, and amplitude.

The sine wave is a good model of the typical wave. It is easy to picture, easy to measure, and similar in appearance to some familiar waves we experience in our daily lives. A study in this wave is a useful beginning exercise.

We can begin to understand waves by looking at the simplest waves first. If you were to tie one end of a rope to a post and move the other end up and down rapidly, it would create a familiar-looking wave. The wave form moves from your hand toward the post. Though the wave appears to move in this direction, any single point on the rope stays put in relation to its neighboring points and in relation to the entire rope itself. No one would suggest that pieces of rope move toward the post and are replaced by newly formed rope.

Ocean waves behave similarly, although they are slightly more difficult to understand intuitively. Standing on a beach, we can see waves moving toward us, cresting, and eventually crashing near the shore. We see a surfer on a board gliding toward us. Everything about this experience suggests forward motion of the wave. However, we also can see another surfer farther from the shore, sitting on his board and bobbing up and down with each wave, moving no closer to the shore with each passing wave.

Even in this simple example, the stationary surfer complicates our intuitive notion of the forward-moving wave.

The standing surfer is riding down the slope of the wave, obeying the rules of gravity. He is clearly using the wave to move forward toward the beach. The wave itself, however, oscillates up and down. Only when the wave reaches the shallows does it fall over on itself, as it can no longer support its own weight. Once the wave crashes near the shore and subsequently causes a splash, it is no longer following wave behavior.

In open water, waves behave as expected. They are, as all waves are, concentrations and rarefactions of elements. When a wave of water moves, the water molecules move up and down. They only move laterally as far as necessary to make the water pile. The wave of concentration moves toward the shore, but each individual molecule remains relatively stationary. When a wave moves in the ocean, its speed can be measured in miles per hour or feet per second. By definition, distance per time is velocity and describes motion. In a perfectly efficient wave moving at five miles per hour, any single water molecule will have moved absolutely nowhere after one complete cycle.

All waves behave similarly to the waves we've just looked at in the rope and ocean examples. This includes waves that we cannot see so easily, such as sound waves. Sound waves differ only in our inability to perceive them quite so easily. When a starter's pistol is fired in the distance, you see a puff of smoke and a second later hear the bang. Sound waves travel from the source, the pistol, to the detector, your ears. Textbooks show us the image of curved lines moving from source to detector. We can easily accept this oversimplification as it roughly fits with our perception. In actuality, an enormous number of compression waves move out in every direction at once. We can measure the wave velocity (the speed of sound is 340.29 meters per second at sea level), wavelength (pitch), and amplitude (loudness). The measurable velocity suggests that the wave has motion.

A sound wave, like all waves, is a compression and rarefaction of elements. In this case the elements are air molecules. Sound is transmitted when a vibration compresses local molecules. But the local molecules do not travel to your ear. These molecules move only as far as the compression energy sends them in vibration. As each molecule moves toward its neighbors, it is in its compression phase; as it moves away from its neighbors, it is in its rarefaction phase. As the vibration oscillates back, it results in a new configuration of compression and rarefaction. As this single molecule moves back and forth, it loses energy with each oscillation until it settles back into its original state. Each of its neighboring molecules experiences the same compression energy and resultant vibration.

So far the sound wave has not moved very far. The key to the wave moving from source to detector (your ear) is the translation of the vibration to the neighboring molecule and then to the next, and so on down the line. The molecules in direct contact with this compressing force move in the same direction with whatever translated energy they receive. No single molecule moves any further than the vibration's energy is able to move it. This movement is enough for it to interact with the next molecule in line and to translate a version of the original vibration; the next molecule then interacts with its neighbors in succession, with an ever-expanding number of members propagating the wave front. This translation occurs molecule to molecule, at the discrete speed of sound.

Sound travels at 340.29 meters per second in air at sea level and standard pressure and temperature. Air density in these conditions is approximately 1.293 grams per liter. Relative molecular weight of an air mixture is approximately 29 grams per mole. This is $2.67 \times 10^{22}$ molecules per liter of air. The average surface area of an eardrum is $5.5 \times 10^{-5}$ square meters. Taken together, all of this means that in a single second, the column of sound waves traveling through a cylinder with the area of an eardrum vibrates $5 \times 10^{22}$ molecules.

It is difficult, using our limited experience, to fathom the incredibly minute size of the molecular vibration and the immense speed of the sound wave. But this pales in comparison to the minuteness and speed associated with the movement of light.

Rope, ocean water, and sound, as well as other forms of natural information, travel in waves, through media. But in the current understanding, there is one holdout to this behavior: light, which, as we know, is explained as exhibiting both wavelike and particle-like behavior. Further, light and sound waves are said to be transverse while mechanical waves are said to be longitudinal. That is to say, as opposed to the peaks and valleys of a typical ocean wave, the compaction phase of a sound wave has more particles smashed into a smaller volume followed by an area where the concentration of air molecules is rarefied.

Possibly, light waves follow a similar pattern, as it is unlikely that nature would choose to make an exception for this one type of wave. It makes sense that light travels as all waves do, though a medium: the luminiferous aether. The aether is an archaic concept, and it is certainly uncomfortable to take such an archaic concept and use it to backpedal into and ultimately undo strongly held present-day beliefs. But many keys to unlock the understanding of the universe can be found in light's wave form.

When we measure light transmission, what exactly are we measuring? What exactly is moving? We are taught that an electromagnetic impulse moves both as a particle (like a bullet shot from a gun) and as a wave. But, we are told, this wave is unlike all of the other waves—the rope, the water, the sound wave—described previously. Rather, this is a moving particle-wave that moves from source to detector, as opposed to the oscillating wave of sound and sea in which none of the wave's components actually travel.

Nature uses waves to transmit information and does so in an organized, predictable fashion. Although we see this repeatedly in every other natural system, we suspect that electromagnetic information follows a completely novel set of rules. To explain this discrepancy, we have invented a dual wave-particle theory of the existence of light. The theoretical gymnastics required to explain these phenomena should be a clue that the prevailing model is very wrong. While the duality model is capable of satisfying equations and experiments, it does not satisfy rational logic. History tells us that Occam's razor should not—and, as this book argues, Occam's razor *does* not—spare waves from its simplifying blade.

As we've seen, waves are physical phenomena of natural movement. Sometimes they fit nicely into our daily sensual experience, like the rope described above, and sometimes they act on a scale so small and fast that they are beyond our perception and are difficult for us to understand. Even on these smallest and fastest of scales, however, waves are still natural waves that obey the rules of waves. They share all the same general properties and laws of physics. We have found that electromagnetic waves are no exception, and I argue that light is no exception, either.

To understand how this is true, we must first return to the seventeenth century, before things got quite so complicated. As we discussed, Christian Huygens proposed a wave theory of light in 1678. A key component of Huygens's proposition was that light is propagated in all directions as a series of waves and travels like all waves do, through a medium. The medium, Huygens proposed, is the luminiferous aether. In the three hundred–plus years that have passed, the notion of the luminiferous aether has been dismissed and is now an archaic concept, most recently because Einstein didn't need it for his theories to work (and because he didn't need it, he dismissed it altogether). If you accept the existence of a luminiferous aether, a medium through which light travels, you can then explain how light travels through a medium, as all other waves do.

But even if you can temporarily suspend your lifelong (century-long) belief that light behaves differently than all other waves, it takes an even bigger leap to buy into the concept that *matter* might share the same "wavy" process as the transmission of electromagnetic waves.

$E = mc^2$. The framework of the entire universe can be accounted for by such a simple equation as $E=mc^2$. Some may read this equation as a formula to change mass into an enormous amount of energy. But the equals sign does not read "can be converted." Rather, this equation states that when you look at matter you are looking at energy, energy in a form that we, in our normal daily experience, are not used to considering. This equation does not mean there is a certain amount of energy that can be harvested from breaking down the bonds of matter. What it means is that matter at its most basic level is energy, no more and no less. It is energy packed in a way that we can hold in our hands. When equating two values, we often will apply a constant so as to keep the results in units with which we are familiar. Einstein's equation could just as easily could read $E = m$. The term "$c^2$" simply allows us to compare energy to mass amounts that we are used to seeing in our common experience.

To show that $E = m$, we could define a new unit of energy equal to the conversion of one gram of ordinary matter. Due to the enormity of the speed of light times itself, one gram would equal $9 \times 10^{20}$ ergs. We could name this new value the Albert, and then one gram of ordinary matter equals one Albert of energy.

Having said all of that, energy is not readily available to us in this form for our use. It is packaged in a way that is far too stable for us to harvest. Though this may sound unfortunate in terms of heating our houses, it is actually what makes matter stable enough that our world does not explode in a fissile event. This book, for instance, has enough energy to wipe out a city.

A simple thought experiment illustrates the equation of matter to energy. If you carry a liter of water from sea level to the top of Mount Everest, it would take 87,000 J of energy. Once up there, you place it in a sealed container, encase it in concrete, and throw away the map. The water is not going anywhere anytime soon. The energy that was imparted into this water to get it to this high elevation will exist in this form well into the future. For all intents and purposes, it is locked away for eternity. Though technically its potential energy persists and is in the amount equal to the enormous effort of carrying it up there, this energy is not readily available for future use. We know it exists, and we can calculate its exact value, yet this energy will remain hidden from us forever. In a similar fashion, most of matter's energy is equally inaccessible to us through common means available today.

This inaccessible energy is all around us. Where does all of this energy come from? There once was an event, or possibly even multiple events, that was (were) characterized by an enormous amount of energy, a Big Bang. This event predated matter as we know it. In the timeline immediately following that event, there was a period of balance between enough energy to package into these supercharged energy packets (matter) but not so much energy and radiation that these configurations were instantaneously shredded apart.

As the cooling process continued, various levels of complexity were obtained to allow for these packets to form into primordial subatomic particles.

The timeline that led to this stable packaging is described in all its detail in Steven Weinberg's book *The First Three Minutes.* Weinberg explains, for instance, the packaging of an electron. Based on the mass of an electron and the equation $E = mc^2$, we can calculate the resting energy of an electron to be 0.511 million electron volts (where an electron volt is the energy required to move one electron through a potential of one volt). This is a very small amount of energy indeed.

The kinetic energy (temperature) for any particle is obtained by dividing its rest energy by Boltzmann's constant (discovered by Planck, not Boltzmann) which is 0.00008617 eV per degree. The approximate temperature, therefore, at which electrons and positrons could exist in a stable manner occurred when the universe cooled to six billion degrees Kelvin.

We can proceed with our daily lives without this knowledge of the origin of matter. We take for granted that matter exists, and that's fine and sufficient for our day-to-day experiences. It certainly is interesting to consider, however. The overwhelmingly enormous scales of the universe's matter supply an interesting puzzle.

As scientists have grappled with this puzzle, the basic players of the "massing" process have been postulated in various forms: luminiferous aether, quantum foam, quantum jitters, the cosmologic constant, Higgs, strings, branes. These terms might describe some of the attributes and functions of elemental units in the process, but does any of them refer to the most basic, essential unit?

A recent discovery covered widely in the news seemed to have an answer. It has long been theorized that a particle is responsible for imparting mass on all other particles. On December 13, 2011, the Higgs boson was believed to be identified in experiments at the CERN laboratory. Since this discovery, the Higgs boson has become widely accepted as the answer to the question of how mass originates. The explanation for how the particle works is that the Higgs boson is the mediator between energy and mass by means of the all-pervasive Higgs field. As particles move through this resistant field, they experience some sort of interaction with the field, or drag. The Higgs boson is responsible for or mediates the field that interacts with the particle, thus giving it mass. A particle does not manifest mass on its own, this model holds; it is only through this interaction with the field that an impression of mass is created. A photon is massless because it does not interact with the Higgs field. Without this field, any proton in our universe would experience no resistance and would be massless.

In 1993, scientists were offered the challenge of coming up with the simplest, most accurate analogy for the Higgs boson. The winning analogy, offered by Professor David Miller of University College London, ended up comparing the Higgs field to the crowded room. A person (particle) who entered the room and moved through the crowd would interact with the people. The people who interacted with the person would give that person "mass." The people interacting with the person at any given time were the equivalent of the Higgs boson, lending that person mass. A person who moved through the room without interacting with the crowd would be massless.

Or picture a submarine moving through the ocean. In the submarine's universe, water is everywhere. It surrounds the submarine and fills every nook and cranny. As the submarine moves, it encounters a succession of different molecules of water but is always enveloped. The resistance this boat encounters would be akin to the Higgs field: a field through which the submarine moves, with which it always interacts.

However, there is a major flaw in this analogy, or any other we might make involving our worldly experience. Though somehow this "resistance" imparts mass, it does not impede a particle's progress. The particle does not require an energy expenditure to overcome this drag, which is very fortunate for the theory. Otherwise, before long, everything would come to a standstill. In the crowded room example, this would mean that the person would require no energy to pass through the throngs of people surrounding him. As we know, moving through such a crowd indeed requires a substantial amount of energy. Eventually, as this room approaches infinite size, we would run out of energy and could no longer move through the "field."

So while the Higgs boson is the most widely accepted explanation of the massing process today, flaws in that theory suggest that other viable alternative should be explored.

But first we should look at just how gargantuan of an undertaking it was to assemble all of the matter that exists today. That is, we should first fully understand just how much energy is required to form even a very small amount of matter. To do this, it is necessary to look to the second law of thermodynamics, entropy. It is easy to imagine the entropy that occurs when a clumsy dog knocks over a stack of pennies on a table. It may be more difficult to see how entropy actually occurs when stacking those pennies in the first place. It would seem that the ordering of this stack is the opposite of entropy. True, we go from a less ordered pile to an ordered stack, but it takes a relatively large amount of energy to stack those pennies compared to the residual potential energy as the stack rests on the table.

Entropy is the tendency toward disorder. If you were to shuffle a deck of playing cards, it is most likely that with each shuffle, they would become more and more disordered. We may find, here or there, little pockets of order, but this would be quite random. We would not expect a large-scale, spontaneous reordering of the card sequence. Indeed, it would be extraordinary to see the cards go from entirely random to the specific numerical and suited order that we would have come out of the original box. Extraordinary, but not impossible. In fact, if you were to shuffle the cards $10^{38}$ times, you could expect to see this occur once. To put this in perspective, if you reshuffled every five seconds, it would take $1.6 \times 10^{31}$ years, excluding breaks. The universe has been here a mere $1.4 \times 10^{10}$ years. To take probability one absurd step further, if you were to shuffle the cards an infinite number of times, you would shuffle a perfect order an infinite number of times, thus showing the difference between a really big number and infinity.

We can apply the concept of entropy to the massing process. We can see entropy in the solar furnace as two hydrogen atoms fuse to form a helium atom plus some debris and lots of energy (light and heat). This feeds further nuclear reactions. It takes enormous amounts of heat (kinetic energy) to allow the fusion process to continue. We cannot, for example, take two hydrogen atoms at room temperature and stick them together. Fusion will always be in the setting of a high-energy event. The reason for this is that a portion of the energy is consumed in the ordering process of making the new atom. It is quite inefficient, and most of the energy dissipates off into space as light and heat.

This inefficiency, coupled with the enormous amount of energy it takes to form even a tiny amount of mass, highlights just how huge the amount of energy was that it took to create the total mass of the universe. First, the universe has an enormous amount of matter, in the range of 10,000,000,000,000,000 grams. Each small amount of matter is formed from a very large amount of energy. Furthermore, this process is extraordinarily inefficient so that only a tiny portion of the needed energy is used to form this matter; the rest is lost in entropy. These statements tell us that the amount of energy needed to create the total mass of the universe was very large—on the scale of $10^{88}$ J, assuming a pretty efficient system, which it almost certainly was not.

The Big Bang is a worthwhile exercise in understanding this scale, because it allows us to think in a framework of a singularity with an instance of change, and this in turn allows for some fantastic calculations and predictions. But no one was there to witness the beginning. The likelihood of a true singularity, where the entirety of the universe is fully contained, is remote. It is likely that there was a singularity-like event where a critical mass/energy formation expanded. It is possible that a single sudden event occurred, but it is also possible that this nuclear furnace resembled what we are more familiar with in a rumbling event over some time.

Beyond questions regarding the nature of light and of the original massing process, we continue to grapple with other fundamental questions about the universe. While widely well-regarded, many of these notions are actually quite absurd. Take, for example, the Copenhagen Interpretation and the Many World Interpretation.

Young's double-slit experiment demonstrated the wavelike behavior of light and contradicted corpuscular nature of photons. In addition to photons, the wavelike behavior also does not fit well with the behavior of massive particles, like electrons and beyond. Niels Bohr began the process of rectifying this problem with what is now referred to as the Copenhagen Interpretation.

Specifically in the case of electrons, it was demonstrated that these can exhibit an interference pattern just like light did. This was interpreted as evidence that particles share the same essential nature as photons and is further support that photons exist as particles. That being the case, a nonwave explanation needed to be invented to account for this odd behavior of particles. The solution was superpositions. In this theory, all particles exist as a wave form. This means that they don't have a specific position in the universe but instead exhibit a probability of being in any position. In a way, they are said to partially exist in all possible positions at once: superposition.

But this is discordant with our everyday experience. So this theory further suggests that the act of observation, especially by a conscious observer, namely, a human, causes it to exist in the place where we observe it. Prior to observation, it exists as a probability in all positions, and instantaneously upon observation, it exists in a sole location where we found it. In this view, the existence of all concrete objects depends on humans to notice them. Prior to conscious living organisms, no objects existed in a specific location but instead shared all locations in a probabilistic manner. Much like the refrigerator light is off when the door is closed and our opening the door causes the light to come on, the Copenhagen Interpretation holds that our observation causes an object to exist in one place.

The mechanism, we are told, occurs instantaneously. So as soon as we see something in the location where we happened to find it, it no longer exists in any of the other positions that it previously superexisted. That might seem fine, philosophically, if we are talking about two specific locations in close proximity, such as two boxes on the same table. It is possible, however, to set up the experiment so that the two positions are light-years away from each other at the time of observation. That would mean that two events would happen simultaneously with a cause and effect etiology that involves instantaneous information transfer across light-years. This violates nature's speed limit of c. This caused Einstein to scratch his head a little.

This superposition superstition is actually still taught and believed and is the cause of great philosophical debates—debates not on whether the silliness is true but rather on the nuts and bolts of its mechanism.

An opposing view, also still widely regarded, is the Many World Interpretation. In this theory, the conscious observer is still in his position of importance. It is he who unlocks the superposition into a final position. He does this by splitting in two. In every act of observation, multiple parallel universes occur so that each superposition now has a universe where that particle exists. In the parent universe, the particle has a probability of existing in Position A and Position B. With the act of finding it in Position A, we have caused a universe of Position A placement and instantaneously created a universe of Position B placement. It is unclear if each observer now exists in each of these separate universes. One would have to presume so. That is an awful lot of universes being created with every observation. Seven billion people observing trillions of microevents per second. Lots of universes, indeed.

I'm not so convinced that either of these represents a reasonable option. We look down on the luminiferous aether and still teach this drivel.

If we are so bold as to begin with the concept that the luminiferous aether is not so far-fetched or at least, in any event, not as far-fetched as the Copenhagen and Many Worlds Interpretations, and that there may be a medium through which light waves travel (as is the case with all other waves), and use that foundational understanding to look anew at the questions facing particle physics today, we are able to find astonishing results. The key to uncovering these results is not just accepting that something akin to the luminiferous aether does indeed exist, but more importantly, looking into what exactly that aether might look like. What is it made of? How does it exist? Once we begin answering those questions, the seemingly disparate subjects of today's particle physics milieu, string theory, the theory of relativity, wave and field theory, discussions of matter, and ultimately discussions of the origin of the universe, can be understood as part of a cohesive theoretical framework. And the concept that holds the framework together is an all-pervasive medium...dare we call it the aether?

# *Chapter 3*

## WARMESTRÄHLUNG

*You cultivate the virgin soil,*
*Where picking flowers was my only toil.*

—*A. Sommerfeld*

*You picked flowers-well, so have I.*
*Let them be, then, combined;*
*Let us exchange our flowers fair,*
*And in the brightest wreath them bind.*

—*Response by Max Planck*

A French military engineer named Sadi Carnot was one of many in the early 1800s looking at engine efficiency.

In particular, the steam-driven piston engine was the state of the art at that time but was very inefficient at converting its source energy into mechanical energy. Carnot's goal was to improve the efficiency and understand the principles that govern an ideal engine. In his only published work, *Reflections on the Motive Power of Fire* (1824), he demonstrated that the efficiency of this idealized engine is dependent on the temperature gradient driving the reaction. This allowed for direct measurements and calculations of the relationship between energy and efficiency and so was the beginning of the specific discipline of thermodynamics.

But if the nascent science of thermodynamics was born in France, it was nurtured and raised in Germany, the center of the physics universe. It might be said that the Friedrich-Wilhelms University in Berlin (Berlin University) was at the epicenter. In the 1840s a German physicist and mathematician, Rudolph Clausius, who had studied in Berlin and elsewhere, was at the forefront in many arenas of the physics world, but today may be best known for his work in thermodynamics. In his paper "On the Moving Force of Heat and the Laws of Heat which may be Deduced Therefrom" (1850), he described the obligatory loss of energy in a heat engine (Carnot's Principle), which contradicts the conservation of energy principle of the first law of thermodynamics. In his 1879 book *The Mechanical Theory of Heat*, Clausius coined the term "entropy" to define, in mathematical terms, this measurable and predictable amount of lost or wasted energy. As we will soon see, Clausius's work, and especially his concept of entropy, would have a transformative effect on one of the most important scientists of the twentieth century, Max Planck.

Physician and physicist Hermann von Helmholtz found his final professorship in physics at the Berlin University in 1871. Helmholtz was one of the best-regarded of all German scientists in the second half of the nineteenth century, "known throughout Europe as a brilliant polymath, moving easily between medicine, biology, and physics" with work on subjects as diverse as "hearing, sight, moving fluids, electromagnetism, and mechanics."[13] Due to his professional stature and influence, in his work on electromagnetism, Helmholtz was certainly a guiding force for the next generation of scientists. Helmholtz was very much a believer in the first law of thermodynamics, the conservation of energy. Considering his influence, Clausius's work on entropy, which would become the second law of thermodynamics, was beginning to cast a new light on the entire field of thermodynamics.

In 1875, Gustav Kirchhoff became chairman of physics at Berlin University, joining Helmholtz. His specific interests were in spectroscopy, thermodynamics, and electromagnetism, and much of his work focused on the measurement of the heat radiation and wave oscillations emitted from objects. Kirchhoff had in 1860 come up with an ideal model of a body that completely absorbs all incidental rays, and neither reflects nor transmits any, proclaiming, "I shall call these bodies perfectly black, or, more briefly, black bodies."[14] A blackbody is an idealized extreme condition used mathematically to define certain physical properties. Theoretically, it is a physical entity that at a constant temperature and pressure absorbs 100 percent of all light striking it. It is perfect at absorption and does not depend on radiation source, type, frequency, angle of incidence. In this theoretical model, it is in radiation equilibrium. While absorbing all radiation striking it, it is able to emit radiation (blackbody radiation). This goes by a made up word called emissivity. A blackbody has an emissivity of 1.

In theory, a blackbody is an ideal emitter. While absorbing all incidental light, it can emit the energy from an internal source. Actually, better stated, as it is in equilibrium, it must emit its internal energy and do so perfectly, completely, diffusely.

Enter Max Karl Ernst Ludwig Planck, born in Kiel, Germany, on April 23, 1858. He was the son of Johann Julius Wilhelm von Planck, a law professor, and Emma Patzig. He obtained a PhD by the age of twenty-one and served as professor of theoretical physics at University of Munich, later succeeding Kirchhoff at Berlin until his retirement in 1926. Planck was awarded Nobel Prize in Physics 1918 and declared the father of quantum theory not long after. In his spare time, while not altering the course of physics forever, he was a mountain climber, pianist, and—as we see in this chapter's epigraph—amateur poet.

In 1889, Planck was offered and accepted a position at the Berlin University, teaching theoretical physics, taking Kirchhoff's place after his death. Though Helmholtz first preferred Heinrich Hertz or Ludwig Boltzmann for the job, both of them passed on the offer, and so began Planck's professorial career in Berlin. Planck became closer to Helmholtz during this long period in Berlin, eventually seeing him as a father figure in his academic life.

Once his academic career finally took off, after several postgraduate years living with his parents, Planck established a reputation as a brilliant physicist and lecturer. He was a leader and twice president of the Kaiser Wilhelm Society. He sat on the review board and later coedited for the prestigious periodical *Annalen der Physik*. It was, in fact, through this journal that he discovered Einstein's work and became his biggest champion. If it weren't for Max Planck's support, the theory of Special Relativity would not have come to such prominent light, or at least not with the speed and widespread acceptance it did. As Brandon Brown writes in his 2015 biography of Planck, "It is no exaggeration to say that Planck discovered Einstein, and he helped shepherd the square peg of Einstein's genius into the round hole of the science establishment."[15] He was a treasured figure in the intellectual world of Germany at that time and is still considered one of the great intellects of the twentieth century.

One cannot fully discuss the multidimensional Planck without touching on his nebulous connection with the Nazi party, as Planck was one of many influential German physicists to live through both World Wars and was thus faced with the resulting moral conundrums. No one alive will ever fully know the extent of his involvement with the Nazi party or of his knowledge of the atrocities that were being committed in his homeland. Present historical accounts suggest that he was an entirely honorable man who was devoted to his country, as were many who lived through the transition from the nineteenth-century loose confederation of German-speaking states into a single German nation and therefore obligated to stand by her as she went through these struggles. Though the precise nature of Planck's involvement with the Nazi party is not clearly understood by present historians—and the same can be said for much of his personal life—Brandon Brown offers a summary in his 2015 book *Planck: Driven by Vision, Broken by War*, which includes the following summary of Planck's place in the complicated historical setting into which he was thrust:

> Planck was 75 when Hitler came to power. He stood with his Fatherland, no matter its warts and crimes, even at the cost of friendships. In summary, Planck dedicated himself to the idea that he could do the most good from within the Reich. His dedication to this notion paralleled his dedication to a homeland that would endure and overcome anything, by it war, famine, or Nazi leadership.[16]

Some facts about this involvement are clearer than others. For instance, it is well-known that he met face-to-face with Hitler in 1933, and that the meeting did not go well. Another indisputable fact is that in 1944, when Planck was in his eighties, his son was ordered to be executed, along with roughly one hundred others, for his alleged involvement in a plot that nearly assassinated Hitler, and Planck worked tirelessly to lessen the severity of the sentence. By this time, Planck had fallen completely out of favor with the Nazi party, and Nazi scientists attacked both Planck's character and his scientific legacy.[17] Despite these questions of his character by Nazi sympathizers and Nazi victims alike, this dubious period of Planck's life came toward the end: he would die just three years after his son's conviction, and most of Planck's most important scientific work was completed decades earlier. Though a full account of Planck's life must cover these later years, it is also important to look to his formative years, when times were much less bleak.

Though he was in many ways a prodigy, Planck's rise to scientific fame, and particularly his years as a graduate student, was not without turbulence. Planck's formal education was in experimental physics, as there were no professorships or classes in theoretical physics at the Universities of Munich or Berlin at that time. Theoretical physics was not yet a highly regarded intellectual pursuit. In Berlin, Planck was a student of both Hermann von Helmholtz and Gustav Kirchhoff, and was inspired by both as he focused his early work on thermodynamics. However, it seemed to Planck that neither of these towering figures had much regard for him early on. As Planck notes in his *Scientific Autobiography*, "Helmholtz probably did not read my paper at all…[and] Kirchhoff expressly disapproved of its contents"[18]. In return, Planck did not feel that these professors "quenched (his) thirst for scientific knowledge."[19]

While Planck would eventually form friendships with Helmholtz and Kirchhoff, it was Rudolf Clausius and his work on entropy that had the most lasting impact on Planck as a young man. Brandon Brown notes that "when Planck read the second edition [of Clausius's *The Mechanical Theory of Heat*], it branded his supple young mind with a permanent new faith." It was this new concept of entropy to which Planck devoted his dissertation. The concept that the entropy of a system could increase while the energy of a system remained the same seemed to be, in Planck's opinion, lost on Helmholtz and Kirchhoff: "None of my professors at the University had any understanding for [my thesis], as I learned for a fact in my conversations with them."[20] Despite the indifference (in the case of Helmholtz) and resistance (in the case of Kirchhoff, from the two most powerful men at his university, Planck moved steadfastly forward with his studies in this new second law of thermodynamics, and indeed, "Entropy became a lifelong lodestone for his thinking."[21]

Planck's interest in entropy was consistent with his stated belief that the noblest and most worthwhile task in science is to search for the absolute truths. His tasks were driven by the hot topics of the day which, in addition to entropy, included the behavior of energy and structure in various systems including dilute solutions and radiation emitters. At the turn of the twentieth century, physicists were stumped by a certain mathematical anomaly caused by taking Maxwell's electromagnetic equations to their logical conclusion. The basic problem was that by applying Maxwell's equations to measure the energy inside a perfectly insulated oven, you would find a seemingly absurd result: that the energy inside the oven, at any temperature, was infinite. This problem is representative of the fact that at as the nineteenth century was coming to a close, there were a number of problems in thermodynamics begging for a theoretical tightening up, and Planck would prove to be the perfect theoretical physicist for the job.

As his career progressed, Planck felt increasingly strongly that the second law of thermodynamics, the law of entropy, was the key to the truth. Many of his contemporaries, most notably Helmholtz, did not hold deeply that entropy was even a true law but saw it more as a tendency or a result of probabilities. This, of course, put Planck's views in direct conflict with Helmholtz and other leaders of the scientific establishment whose higher academic positions at the time allowed them to dismiss Planck's early contributions as useless. In an experience common to contrarians within the academy then, as it is to this day, Planck had to fight against the establishment of egos and self-promotion to establish his work and the new direction in which it was leading.

Boltzmann, an outspoken figure and a force of nature himself, was a severe critic early on. Planck had criticized the value of Boltzmann's results in the field of gas theory, whereas Boltzmann condemned the irreversible nature of entropy. Although the arguments went back and forth for some time, sometimes getting quite contentious, in the end Boltzmann acquiesced to the formulations of Planck, and in time the two would come to respect each other's work a great deal, even if they never became fast friends. In fact, it may have turned out that Boltzmann's arguments and mathematical tools allowed

Planck to refine and re-explore the initial ideas, which would eventually result in Planck's most important discoveries, most notably his namesake quantum value, which would forever alter the course of physics.

For Planck, the blackbody absorption was problematic, and he realized that it was impossible to have a body that was accepting all radiation without reflecting at least a little at the surface, possess the thinness required to have this one-way characteristic, and contain the radiation so that it would not bounce back out. It was the emitted light that gave him insight into the properties of radiation. His calculations on blackbody phenomena allowed him to surmise that radiation is an electromagnetic phenomenon.

As Planck struggled with his claim of a resonator exerting an irreversible action on the energy of the surrounding radiation field, he turned to thermodynamics. Here he employed his beloved second law to bring the entropy rather than the temperature into relation with its energy, later remarking in his Nobel lecture that "from the start I tried to get a connection, not between the temperature but rather the entropy of the resonator and its energy, and in fact, not its entropy exactly but the second derivative with respect to the energy since this has a direct physical meaning for the irreversibility of the energy exchange between resonator and radiation." After this initial discovery, and after what Planck called "some weeks of the most strenuous work of my life,"[22] he would make another, even more profound discovery. In short, Planck was able to determine that the energy of an emitted "ray" is proportional to its frequency. It is proportional to an amazingly precise and universal degree. This proportionality constant he designated *h*.

Planck's conclusions would ultimately go on to forever alter the course of physics and form the foundation of quantum mechanics. In his own words, Planck says, "Radiant heat is not a continuous flow and indefinitely divisible...It must be defined as a discontinuous mass, made up of units, all of which are similar to one another."[23] Though Planck did not immediately accept its wide-ranging implications, the idea of the energy quantum would become the foundation of quantum mechanics. Indeed, Einstein saw the breadth of the implications even before Planck did, applying Planck's "packets" to the realm of light in his solution to the photoelectric effect, which Planck initially resisted as an overreach of the quantum's bounds. This reluctance was not atypical of Planck for, as Brown notes, Planck could be considered "a revolutionary who brought quantum to the world, even if he has only reluctantly played along after his first inspiration."[24] Eventually, of course, both Planck himself and the scientific establishment (beyond Einstein) began to comprehend the monumental importance of the quanta beyond the context of Planck's initial work on blackbody radiation. Written nearly fifty years after Planck's initial discoveries of the energy quanta in blackbody radiation and when the profundity of the implications were clearer, Planck's *New York Times* obituary notes that while Einstein's special and general relativity "has given man a new and more profound understanding of space and time, matter and motion," quantum mechanics "provided man with a master key to the universe within the atom, to the nature of light and radiant energy in general."[25] And if quantum theory provides a master key to the universe at the smallest subatomic scales, then *h* provides a master key to quantum theory.

Planck's constant is truly the fundamental value of the universe: more fundamental, more primal than *c*. In fact, as I will show later, the speed of light is totally dependent on *h*. The value of Planck constant is the most descriptive element of the universe.

Planck's own work that led to the calculation of his namesake constant was obviously the most groundbreaking and important of his findings, but contributions of other scientists in the years leading up to Planck's work at the turn of the century are worth discussing. In the 1880s, Johann Jakob Balmer, the Swiss mathematician who also studied at Berlin University, was studying the spectral lines of hydrogen. Small amounts of light are emitted from elements as their electrons move from a more excited state to a less excited state. These levels of excitation are very specific and occur in discrete steps. Spectroscopy shows these refracted lines in a pattern that can be measured directly. Balmer used the values previously published by Angstrom to predict the wavelength of each line. He fashioned an empiric mathematical formula that at least for the specific case of hydrogen, which would account for these stepped wavelengths. From a historic standpoint, it was these spectral lines that are used to classify and otherwise study stars.

Roughly contemporary to Balmer, Johannes Rydberg, a Swedish physicist, was working on (struggling with) defining the spectral properties of other chemical elements. He specifically was working on alkali metals when he thought to apply Balmer's work on hydrogen atoms. Using Balmer's equation, he was able to create his own equation that could be more generalized and apply to all elements and their emitted light. The original Rydberg formula for hydrogen was:

$$1 = R(1n21 - 1n22)$$

This was later modified to:

$$E_n = -hc_0R_\infty/n^2$$

where $E_n$ is the quantum energy for each excitation state, and $R_\infty$ is a constant derived in accounting for wave-numbers of spectral lines translatable to all atomic series. This work done by Balmer and then Rydberg was an early hint at quantum theory.

In 1896, Wilhelm Wien, Planck's colleague and eventual coeditor of *Annalen der Physik*—who, incidentally, had worked under Helmholtz at the Berlin University—empirically determined the distribution behavior of radiant energy. Wien was a prolific scientist who would eventually isolate a positive particle in the hydrogen nucleus that would later be identified as the proton, but in 1896, his key contribution to science was his discovery that the energy of a blackbody signal was proportional to the wavelength of that radiation: that is, the coefficient *R* in the aforementioned Rydberg formula is dependent on the wave number (frequency) and therefore is inversely proportional to the energy. Wien further showed that this energy can be correlated to temperature and that wavelength is therefore directly proportional to temperature. This is known as the Displacement Law of Wien, though it is worth noting that "most physicists accepted Wien's equation as an empirical approximation to the truth, if not the gospel truth."[26] Empirical observations in later years would show that Wien's equation was indeed an approximation, and it would be left to other scientists to sharpen this approximation into a finer point.

As Wien made his discovery, Planck was also at the Berlin University, by this time having become a full professor. Planck was dear friends and colleague to "Willy" Wien, and in a sense, Planck and Wien could be considered to be working to fill the vacuum left in Berlin by the 1894 death of Helmholtz. Planck built on the empirical work of Wien and developed a theoretical framework to explain the relationship between wavelength and energy. This fit the specific data under investigation and became known as the Wien-Planck law.

By this time, Planck was editor of *Annalen der Physik* and was receiving frequent papers in direct contradiction to Wien's and his work. In Planck's own words recounting those days:

> Under more scrutiny a problem arose, this was not a universal law but only fit satisfactorily for small wavelengths. Otto Lummer and Ernst Pringsheim found considerable deviations were obtained with longer wavelengths and on the other hand measurements carried out by Heinrich Rubens and Ferdinand Kurlbaum with infrared residual rays disclosed a totally different,

> but under certain circumstances, a very simple relation characterized by the proportionality of the value of R not to the energy but to the square of the energy. The longer the wavelength and the greater the energy the more accurately did this relation hold.[27]

Max was indeed frustrated by this. His work to define the energy of blackbody radiation was shown to no longer be universally true. At approximately this same time, his father died in Munich, foreshadowing the many tragic personal losses that would pile up for Planck over the next four decades. Showing his characteristic strength and stoicism, Planck continued to be productive in his professional life, searching his thoughts about the available data and the nature of radiant energy until the solution struck him:

> Thus two simple limits were established by direct observation for the function of R: for small energies proportionality to the energy, for large energies proportionality to the square of the energy. Nothing therefore seemed simpler than to put the general case R equal to the sum of a term proportional to the first power and another proportional to the square of the energy, so that the first term is relevant for small energies and the second for large energies; and thus was found a new radiation formula.[28]

This explains that energy and wavelength are related but still does not yet predict the stepwise fashion by which these energies are manifest. Classical physics at the time would suggest that energies as well as all natural occurrences could be measured as a continuum. As it turns out, Planck found that when directly measuring the energies of radiant emissions, there are very specific peaks. The wavelengths of the spectral display for an element occur at very specific points and these are related to each other by this proportionality relationship. In this relationship, the proportionality is related to a constant *h* so that when looking at these wavelength peaks only certain energies emerge. These energies are all multiples of a basic number. No energy is seen below this number; no energies are seen as fractions of this number. By applying this equation, it became evident that the world existed in discrete steps and not the slur that we once assumed. Planck's "new 'essential step' was a radical and subversive one, the first time that anyone had considered treating sacrosanct *energy* in a gritty way."[29] This "grittiness" is what Einstein would attribute to light in his working out of the photoelectric effect. It is the foundation of quantum physics. As Planck himself speculated privately in 1908, this grittiness could be applied to time itself and even to the then-brand-new concept of space-time put forth by Einstein and Minkowski.[30] This hypothesis has not yet been proven, and we will discuss it further in this book's subsequent chapters.

As is true of all great scientific advances, Planck's discovery rests on the existing body of work at the time. In Planck's case, much of the groundwork was laid by Balmer, Rydberg, Boltzmann, and Wien. Planck famously quipped, in what is now often referred to as Planck's Principle, that "[a] new scientific truth does not triumph by convincing its opponents and making them see the light, but rather because its opponents eventually die, and a new generation grows up that is familiar with it."[31] Ironically, it took some convincing for Planck himself to fully see the light, as he became part of the "old guard" that defended the more radical interpretation of the quanta's implications as put forth by Einstein and the younger generation. For example, it was several years until Planck accepted Einstein's work on the photoelectric effect, which applied Planck's energy quanta to light. But by the 1920s, the quantum revolution was fully underway.

The integral piece of Planck's legacy is the knowledge that our universe can be described in small discrete steps, or "quanta." These quanta are so incredibly small as to appear as a continuous field. They are so small as to be undetectable by even our most powerful tools. But they are clearly evident in the behavior of electromagnetism and beyond. We have contrived ideas of small size, small distance, and small mass, but our simple notions cannot begin to capture the infinitesimal smallness of *h*. And in the smallness of this number is what remains locked up as the entirety of the universe. It is the link between small and large.

In his autobiography, Planck mentions, "It is one of my most painful experiences of my entire scientific life that I have seldom, in fact never, succeeded in gaining universal recognition for a new result, the truth of which I could demonstrate by a conclusive, albeit only theoretical proof." Planck's contributions to science—which, as the above quote indicates, were primarily theoretical rather than experimental—are immense. As we've already discussed, Planck's earliest interests were on thermodynamics, thermoelectric effects, entropy, and the theory of dilute solutions. His work on radiation, however, advanced science in general and quantum physics specifically at its most basic level. For this he gets a couple paragraphs in most science books and has that letter *h*. Indeed, the essence of what is taught about Planck today is that "[h]e discovered quantum theory. He had a mustache. And that's about it."[32] This hardly captures the true importance of his findings, which are comparable to those of Newton and Einstein.

A few years after defining the Planck constant, Niels Bohr recognized that this stepwise quantification of the universe was especially applicable to atomic structure. While his model proved to be an oversimplification, his electron orbits still have merit in how their energy states exist in steps exactly defined by the Planck constant.

These quantum steps are related throughout our known universe by this one universal constant, *h*. In other words, scrutiny of all objects and events from using a flashlight, a child swinging from a tree, making ice in your kitchen, or measuring a quark end up at their most critical level depending on one single common denominator: the Planck constant.

Its numeric value depends on the units with which one chooses to express it. For instance, it is commonly referred to as $6.62606957 \times 10^{-34}$ $m^2$ kg/s or $4.135667516 \times 10^{-15}$ eV·s. It is an amazingly small value.

It is taught to most of us as a constant, a resonator, a number that links two important values by a degree of proportionality. Let us now consider this value as more important than just its linking ability. We need to explore this value as if it might have a specific property of its own.

The term "resonator" is very interesting. What specifically is resonating? Is there some specific entity that is vibrating? In defining the relationship between energy and frequency, one might expect all possible values could exist. More or less energy could be applied to a system in an evenly graduated fashion with an infinite number of possibilities. This is almost true. It turns out that the possible values occur in a stepwise fashion, but these steps are so minuscule as to appear almost as a spectrum. These steps are defined exactly by the Planck constant. Physicists today recognize the existence of the numerical value, but the idea that these represent a *physical entity* has not yet been seriously considered. I argue that these steps are more than a mere numerical value and do in fact represent an actual physical space.

The Planck constant defines the physical space through which energy is transmitted. It has a specific distance, which we estimate as Planck length. Everything passes through this space at a specific rate, which is the speed of light.

Using these parameters and integrating some other derived constants, a host of universal natural values emerge, some more valid than others. (Here I make a personal plea. These values should go by the name "universal natural values" instead of "Planck units" because the term "Planck unit" should be reserved for something very specific, to be described shortly.)

**Planck Constant:** 6.62606957 × $10^{-34}$ $m^2kg/s$ or 4.135667516 × $10^{-15}$ eV·s

**Planck Length:** $l_P = \sqrt{hG/c^3}$ or 1.616199 × $10^{-35}$ m

**Planck Time:** $t_P = l_P/c = h/m_Pc^2 = \sqrt{hG/c^5} = 5.39106 \times 10^{-44}$ seconds

**Planck Area:** $l_P^2 = hG/c^3$ or 2.61223 × $10^{-70}m^2$

**Planck Volume:** $l_P^3 = (hG/c^3)^{2/3} = \sqrt{(hG)3/c^9}$ or 4.22419 × $10^{-105}m^3$

In addition, there are derived values for Planck energy, power, force, density, momentum, angular frequency, current, and others.

The constant *h* defines one quantum. It is the relationship between the speed of light and the smallest unit of time. It is a number that helps define a distance. It is the smallest possible distance. It is the very limit between something and nothing: the distance of one quantum, the distance it takes light to travel in the smallest moment. It is the diameter of a speck of aether.

It is this speck that makes up the entire universe. These specks fill all space. There is no space in between specks. No part of the universe is without specks. Specks are all the same size. The shape of a speck, however, is outside of our common understanding and experience. It must have spherelike qualities because it is Planck length in every orientation. That would define a diameter and therefore define a sphere. The sphere is an incomplete model because, even tightly packed together, there would have spaces in between them. It has cubelike qualities in that, as there can be no space in between these units, they would be similar to stacked cubes. Cubes, however, would have dimensions that varied depending on orientation. For instance, straight across would be shorter than corner to corner.

This Planck unit is a point. It is the singularity of the smallest possible dimension. It is the closest thing possible to dimensionless. It is absolute zero in spatial dimensions. Even so, it is measurable as Planck length by how it behaves.

In building this mental model, one might picture these Planck units all lined up claiming their particular position in space. Maybe we picture these units behave as if they are glued together. This is not necessarily so. Though there is no space between them, they don't necessarily need to be physically linked to each other. Conceivably, they could slide around each other in a random fashion. Things that happen through them occur so quickly that they seem to be a static mesh substrate. This should be clearer in chapters to come, as we use the concept of the Planck unit to tie together the seemingly disparate threads of theoretical physics.

Certain theories come close to recognizing the aetherlike existence of Planck units but fail to make the final, conclusive leap. Loop quantum gravity is a theory that defines this space. It is said to be a spin network that exists in a bigger mesh of spin foam. In this form, it seems to have a dynamic existence where there are things going on beneath the Planck level to give this unit character. In this iteration, there is a granular atomization of the universe that is not just a background but the actual participants in "events" such as particle formation. Across the board, LQG theorists despise the idea of aether while they pretty nearly define the aether's existence. This simple detail may be what is blocking this theory from the Theory of Everything.

# *Chapter 4*

## CHRONOS

We all have a personal experience of time. We have an innate sense of what time means. There is not a thing in our lives that is not inexorably enmeshed in time. Time is the measuring stick by which we meter our lives. Time is one of our most valuable commodities. Even though our lives are so dependent on the workings of time, we really have difficulty defining it. Indeed, it is almost impossible to define time without using some measure of or mention of time's components.

Time, distance, and speed are all intimately related. Time only makes sense as it measures something doing something (i.e., time measures movement). Time can be seen to exist in two very opposing ways. There is the one-way arrow of time that we are so sure of in our personal experience, and there is the time of the events of the universe as defined by the intricate equations of physics. The latter, becoming significant with special relativity in the early twentieth century, seems to behave in an entirely different and counterintuitive manner from the former. In our limited world, we experience events that occur at a relatively slow speed. For the rest of the universe, however, at both the smallest and largest levels, the levels where the laws of quantum mechanics and Einstein's special (and general) relativity are more apparent, everything occurs at or near the speed of light. Such speeds are not familiar to us. We experience our existence through our daily routine. We have an intuitive sense of what to expect, and we have memories of past events and records of others' past events to help guide our intuition. Classic mechanics are obedient at these everyday rates, but as things speed up a little closer to the speed of light (the laws of special relativity, for instance) they seem to change shape and time frame and even seem to have the potential to exist in more than one place at a time.

From our earthly, human experience of time, we know with complete assurance when we go to bed tonight that in seven or eight hours, a new day will begin. We can clearly remember the events of our lives and confidently chronicle them in their particular sequence. The people in our lives will agree with the telling of this story line. We make plans for the future because we live with little doubt that the future is coming.

Our personal experience of the arrow of time does not jive with the mathematical notion in special relativity that no time has any particular immediacy. Special Relativity tells us that there is no such thing as "true simultaneity"—an observer's experience of time is relative to the observer's speed of motion through space. In some models, what we sense as past, present, and future are all immediately available at any given time. This would indeed imply that all of our personal actions are predetermined. It would imply that what we think of as spontaneous choices are sealed in time and not chosen at all. It would imply that we could climb into a modified DeLorean and travel to that past or future time.

If we examine time in the context of Einstein's relativity, however, there is no specific time that is special. Past, present, and future are variables that depend entirely on the observer's viewpoint. Any other observer will view someone else's past as his or her present. No one can claim a preferred position in the universe that would make his or her experience of now any more valid than any other observer, even though both observers are experiencing very different "nows." The difference is measurable to the degree that they are separated by distance and measured by the speed of information transmission, *c*.

From a scientific standpoint, time is a measure that holds a place of great importance, but it is not so clear that it has directionality. That is to say, there is no reason in relativity that time moves in a forward direction with an anticipated interval. As a placeholder in equations, there is no reason for it not to move "backward" as easily as "forward." "That the universe moves irreversibly in one direction is accepted, yet still mysterious," Brandon Brown writes, and he goes on to add that even today, "[s]ome leading physicists suggest we might be well served by focusing more of our attention here, despite thousands of years of head-scratching."[33] Again, we are left with the "weirdness" of time's behavior that has been mathematically proven by Einstein and experimentally verified time and again, but we also have the unidirectional "forward" arrow of time that we experience and that Brown suggests we look into more closely. For the rest of this chapter, we'll do just that.

Can we remember the future? Perhaps time's seemingly forward arrow potentially has to do with our recollection and egocentric viewpoint.

Kurt Vonnegut's *Slaughterhouse Five* was an interesting peek at time. Part of the story occurred at our normal pace. Part of the story explored time through the eyes of beings who saw time's fourth dimension just as we see our customary three: as a continuum, all at once. To these beings, the Tralfamadorians, a human would look like a millipede with baby legs at one end and old man legs at the other. Your entire existence, your time dimension, would be seen in one glance.

In a 2002 article in *Scientific American* called "That Mysterious Flow," Paul Davies explores a universe where all four dimensions are laid out just as the Tralfamadorians would see them. We can pretty easily picture *X*, *Y*, and *Z* axes even as drawn graphically on a two-dimensional sheet of paper. Adding another axis proves very difficult to represent graphically and is equally difficult to picture mentally. The accepted attempt at this is to represent space and time as a block, and thus the model is given the name of Block Universe.

In this theory, time holds a position much more similar to the other familiar spatial dimensions. That is to say that just as every place in the universe exists right now, so does every time. There is nothing special about now, and all of the past and all of the future exist always. All times are now, and any specific time (event) is accessible now to an observer in the right position.

Davies and others confidently describes the Block Universe as the only theory that makes sense. For them, our intuitive flow of time is an illusion.

In the cyclic universe model, there are infinite recurring Big Bangs and Big Crunches. These would have enough randomness and sloppiness so that each new cycle creates a unique and random universe. The more science fiction–oriented view suggests the idea that all the minutest events recur exactly as played out previously and then again in reverse an infinite amount of times.

The difference between the scales of our common experience and the small scales of particle physics and immense scales of astrophysics are ripe for confusion and in fact are the basis of our most enjoyable science fiction, such as time travel.

H. G. Wells's *The Time Machine* was the first innovative view of time as a fourth dimension. While mainly a commentary on the human condition, it brilliantly used this evolving theory to explore the physics of time.

If time could go in reverse or back and forth in any order, it is true that we would most likely not be equipped to sense it. No matter what the order, it would feel natural to us and would always seem to moving in the right direction. Conceivably, we could presently be in the process of a big crunch moving backward in time in a contracting universe, but because we are only able to interpret our forward arrow, our calculations falsely suggest an expanding universe.

If we could vary our time position and rewrite our story, the change and the ramifications would be enormous. As the documentary *Back to the Future* tells us, one change, and your brother's head disappears. Or, as suggested in Block Universe theory, all of time is already laid down and there is no rewrite possible: predestiny. Time travel would be pointless if you could only end up exactly where you were at any given time in the exact same conditions with no knowledge of the travel.

Einstein has taught us many things, two of which are particularly interesting to the present topic of our experience of time. These are time dilation, from special relativity, and the concept of space-time, from general relativity.

1. Time lengthens between two observers in relative motion.

Two objects or observers in relative motion to each other will experience their own individual normal events but an altered view of the other object. Time passes more slowly for an observer in motion. Specifically, mass will appear greater, distances will appear shorter in the direction of travel, and time will lengthen. These differences are more than theoretical and have been demonstrated experimentally (with extremely precise instruments on satellites, for instance). At our everyday walkabout pace, these differences are not measurable, much less noticeable, but according to the equation $1/1\text{-}v^2/c^2$, as an object's speed approaches that of light (in the equation, as $v$ approaches $c$), its movement through time will appear slower to a reference observer.

At approximately 0.85 times the speed of light, time will appear to move half as fast for the reference observer. An example told in many books says that someone leaving Earth in a 0.85 warp-speed spaceship on a twenty-five-year journey will return twenty-five years older while the family he or she left behind will have aged fifty years. How exactly does this work?

The key concept underlying the truth that time passes more slowly for someone in motion than it does for someone who is stationary is the constancy of the speed of light. A "photon clock" is often used to illustrate how an object in motion will experience time slower than an object at relative rest. In this photon clock, a photon bounces back and forth between two mirrors at a regular rate. The clock "ticks" as a photon completes one round-trip from one plate to the other and back. In this clock, the plates are separated by such a distance so that one second takes one billion ticks. Imagine looking at one relatively stationary photon clock while another slides by at a constant velocity. As we should not be surprised to find out, the passage of time is slowed for the moving clock in comparison to the stationary clock. This is because, as the figure below illustrates, for the clock in motion, the photon travels a further distance in one tick: B1 and B2, the distances traveled by a photon in each tick of the moving clock, are each longer than the distances traveled by the photon in two ticks of the

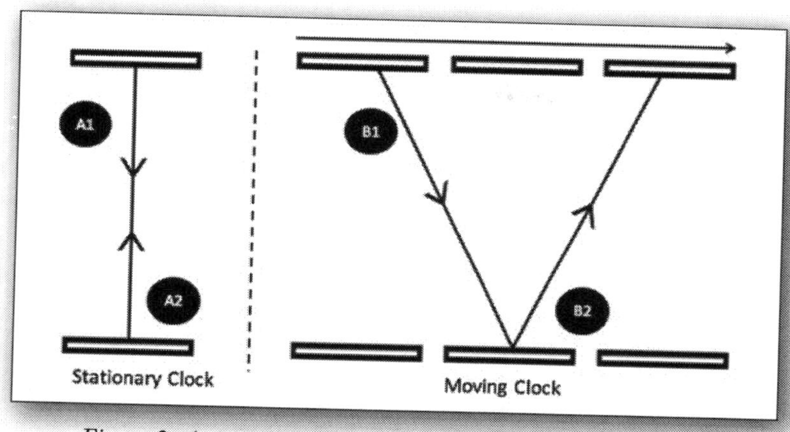


*Figure 2: A stationary photon clock and a moving one, with the photon traveling a further distance in one tick in the moving clock (B1 + B2) than in the stationary clock. Since the speed of light is constant, the tick of the moving clock is longer: time slows.*

stationary clock (A1 and A2). Since the speed of light is constant, the tick of the moving must then take longer. The closer the movement of the clock gets to the speed of light, the longer the photon in the moving clock must travel, and the slower time will appear to move.

The photon clock example explains the mechanism whereby time elapses more slowly for the object in motion relative to the stationary object. Hence the observer traveling near warp speed appears to the relatively stationary observer as having aged only half as much as the stationary observer. But in this case, shouldn't the person in motion be able to live nearly forever? The answer is both yes and no. From the perspective of the stationary observer, his traveling friend could live many hundreds or thousands of years on the stationary observer's speed of time. But, just as important, from the stationary observer's perspective, it's as if the moving person is living his life in slow motion.

In short, Einstein's work on Special Relativity showed us that the constancy of light allows for some behavior of the movement of time, which may be humanly counterintuitive but is perfectly normal for the rest of the universe.

2. Space and time are enmeshed as a single entity, space-time.

Our commonly experienced measured distances can be related to space-time by the equation $s^2 = (ct)^2 - x^2$ where $s$ is space-time "distance", $c$ is the speed of light, $t$ is time and $x$ is the spatial distance (our usual idea of distance) traveled by the observer. The term $x^2$ is negative in this equation, so as an observer covers more spatial distance, his distance in space-time gets smaller; and $c$, the speed of light, is a constant. So less time passes for the observer in relative motion. In relativity, time can be stretched and distances shortened by motion, but all observers agree on this distance $s$: space-time is not relative; it is invariant.

Furthermore, if somehow we were not moving in space (regular distance) then $x = 0$, so $s = ct$. This is completely counterintuitive, as is most of physics. Only by standing perfectly still are we moving through space-time at the speed of light. Everything is moving through space and time at the combined speed of light. We may have difficulty sensing this, but it is Einstein's proclamation that when all four dimensions are considered together, there is one sum result, one fixed space-time speed. As your speed through space approaches the speed of light, your speed through time slows. This is the defining concept of time dilation.

If something is moving through space relative to you (speed through space), then it is moving slower in time than you (speed through time). At the speed extreme of this model, a photon moving through space at a rate of $c$ does not move through time at all. Everything is moving through space-time at the speed of light.

Time dilation and space contraction are accepted as peculiarities of special relativity, but they are much more than mere peculiarities; they are expected and calculable behaviors. For objects at a distance, one object's experience of another object is limited by the time it takes for information to pass between them. This occurs at the speed of light. For two Planck units next to each other, this is $10^{-44}$ seconds. For us to feel the warmth of the Sun, it is about eight minutes.

If a single Planck unit had some sort of self-awareness, it would not experience itself in space-time. Space-time is how that index unit experiences all other Planck units. It takes both space and time for that to happen. In a sense, **space-time is how a Planck unit experiences an event *n* at *x* Planck time ago**. Space-time is always a measure of the past.

Let's look at space-time from the perspective of a single quark. I will discuss the essence of matter more fully in chapter 8, but for the moment we can picture this basic particle making its way through space. In doing so, Planck units are gathered into the quark space, wind their way through the maze of particle folds, and emerge from the other side. This evolution maintains the existence of this particle. If a particle were to speed up, it would pass through Planck units at an increasing rate. It spends less and less time in each Planck station. If at one velocity, it spent one millisecond having a single Planck run its course through the entire particle, then at twice the speed the particle would have entered and departed that Planck space in half a millisecond.

As an aside, and hopefully not too distracting of one, this may be one important facet to the mechanism of the Heisenberg Uncertainty Principle. Our sample quark exists in its present Planck configuration with each Planck unit exactly where it is now for only the space-time equivalent of the speed of light, and that information arrives to the observer at that same speed of light and so is obsolete many times over by the time it is evaluated. Furthermore, nothing can be truly stationary in space-time, or we could know its position and velocity with certainty.

Physicists are confident in the relativity of time. On the other hand, the rest of us are equally confident that we experience the flow of time personally. Is it possible that our collective intuition is nonsense, or is there a way to reconcile these vastly opposing views?

A short digression: Humankind has used numbers for as long as we have had reasonably developed communication.

Once we started naming things, we had to figure out a way to say how many of them there were. More complex number use has been dated back through several ancient cultures. For instance, Sumerians engraved their symbols into clay. There are studies of Akkadians and Babylonians, whose numerical systems have been shown to have been especially utilized in commerce. Amounts of things made intuitive sense, but the absence of things was much more abstract. The Babylonians, for instance, had no symbol for zero. They had a base-60 system, but this would be equally problematic in any system that would require a zero as a placeholder (10, 100, etc.).

Egyptians used a base-10 system and got around the placeholder issue by having a different symbol for each power of 10.

The word "zero" is most likely derived from the Italian mathematician Leonardo de Pisa (ca. 1170–1240), also better known as Fibonacci. According to Alfred S. Posamentier and Bernd Thaller in their book *Numbers: Their Tales, Types, and Treasures*, Fibonacci used the word "*cephirum*," which later turned into the Italian word "*zefire*," which later became the French "*zéro*" and then the English word "zero."

Zero as an absence of anything was a long time in coming and not completely intuitive. Modern man has grown up with the idea of zero. It is a close acquaintance, and so today we feel that we have a good handle on what it means.

For instance, in measuring temperature, there is an absolute zero: –273.15° C. Even so, one could imagine a lower number, say, –280° C. Even though we can imagine it and assign a value, there is no possibility in our universe for it to exist. All motion stops at the absolute lower temperature limit.

It is possible that time has an absolute lower limit that is not zero, though so small, so fast it seems zero-like. This number has actually been calculated or, rather, estimated as the calculations are based on some assumptions. It has been dubbed Planck time. It is the time it takes to travel at the absolute highest speed, *c*, for the absolute shortest distance, Planck length. It is postulated to be $5.39 \times 10^{-44}$ seconds. This is the shortest possible time. This is the absolute-zero equivalent of time.

It is said that when Einstein was a boy, he was intrigued by the idea of traveling at the speed of light next to a beam of light so as to see it in freeze position or another variation, to look at his reflection in a mirror while traveling at the speed of light. According to Newtonian mechanics, two things traveling at the same speed and direction should appear stationary relative to one another. This is not the case for light, which seems to be a violation of classic mechanics. No matter the observer's point of reference, direction or speed of motion, he or she will always experience light traveling at the same speed. This is one of the foundations of special relativity.

Light transmits through the Planck ocean, affecting each individual unit in its path. For our purposes we will call this effect a vibration. This vibration moves unit to unit at the speed of light. This is the obligatory speed—no faster, no slower—and is dependent on the units' uniform Planck size and their uniform intrinsic harmonic. This vibration travels unit to unit until reaching that ultimate tiniest space just before your eyeball. It moves from that Planck unit to your eyeball in the shortest possible time interval, Planck time.

This is time's absolute zero. It is the definition of instantaneous. If one were to try to detect velocity at that single Planck unit, one would run into problems with time's "zero" in the denominator.

The point of all this is that if a light detector is stationary, it should detect light approaching it at the full speed of light. This makes perfect sense. Unfortunately, nothing in the universe is stationary. Everything is relative in position and motion in the grand scaffolding of luminiferous aether. Everything should have a vector to its velocity added or subtracted from the speed of light. Again, this turns out not to be the case.

Every detector, from your eyeball to sophisticated lab equipment, is moving in some direction or another. Each is subject to that moment of detection. That moment happens at Planck time, which is absolute zero time. Instantaneous time plays a trick on the detector. At this instantaneous moment, the detector is essentially motionless in space. No matter what was happening prior to or immediately after this single Planck unit detection, for that instant, the detector is at a standstill. Movement has a speed, and as we are talking about no distance over no time, there can be no measurement of motion. At the absolutely most discrete distance and moment in time, the universe is at a standstill. What we know and expect of our everyday Newtonian world breaks down at the absolute limits of time and distance.

Imagine for a moment a motion picture, an action movie. Cars and jets and bullets are whizzing around at high speed. These all seem to move in a continuous fashion; however, the motion picture is actually a series of still pictures. Each Planck unit is the single still frame of the motion picture of light.

The old high-school philosophical exercise of "if a tree falls in the woods" is a simple attempt at thinking how the world is defined by our experience of it. Time happens in this way. It seems to be operator-dependent. The arrow of time that we are so certain of may depend on our unique position to its occurrence. We witness a beautiful, peaceful, starry night and are pleased to be "in the moment," but much of that starlight actually happened millions of years ago.

We like seconds, minutes, hours, days, weeks, and years. Our brains don't really fathom the meaning of either small extremes of Planck time or the very large extreme of light-years. An atom in your hair and one in your toe may seem far away in your own personal space, hundreds of trillions of atoms apart in fact, but on an astronomic scale, they are essentially the same point.

In contrast to our position on the astronomic scale, the electromagnetic world by which our information comes to us occurs on a scale equally as small. A light "wave" only spent Planck time exactly with each Planck unit at a specific locus, the event was comprised of that specific unit in that exact pattern for $5.39 \times 10^{-44}$ seconds. One tick later, and the electromagnetic process has moved exactly one unit down the line. All electromagnetic events of the universe would have moved exactly one unit. This is the universal standard clock by which our common events are measured.

A sample particle, a quark for instance, sits in the grand Planck ocean we call the universe. It makes its way through this meshwork and does so at a certain rate. A faster moving particle will pass by any specific region in less time than a slower-moving particle. Here is where space-time plays a little trick on us. If one looks at our original particle by the percentage of time that it exists at that aether station, the faster quark may have spent 50 percent less of its existence there than the slower quark, but the transmission of all information travels at the speed of light: no faster, no slower.

It is said that time cannot be described in absolute terms because there is no external reference metric. Well, as it turns out, time does have a definition. The definition is at the extremes of the universe. One tick is the passage of an electromagnetic event across the diameter of a Planck unit, one tick of the Universal Clock.

We are familiar with the ticking sound made by an old-fashioned mechanical clocks. This comes from an internal mechanism called the escapement. It is a device that rocks back and forth over a toothed wheel. It allows a controlled release of energy at a regular rate. As it rocks, it allows the wheel to advance one notch. This is the tick of a watch. The escapement for the universal clock is the travel of one vibration in one Planck unit.

All Planck units are siblings. It is because they are related that they follow Einstein's rules of relativity (pun intended).

At the single quantum level, a Planck unit exists in time and place. It experiences its arrow of time in a very exact manner. It is only at the Planck level that the true arrow of time exists. It is the arrow of space-time and it is particular, not relative.

Relativity does not exist at the single quantum level. Relativity is the way one Planck unit experiences another Planck unit that happens to be going through its own particular space-time. One unit will experience another unit limited by the speed at which information can travel between them: the speed of light in space-time.

We personally feel like we experience an arrow of time because we take up only a small space in the universe. The time it takes for information to go from one end of us to the other is imperceptible to us. From our limited standpoint, our Planck units go through space-time as a single entity. In fact, because of the speed of space-time, not only do we personally feel like we own our arrow, but we drag all of our surroundings and people with us.

In actuality, even two units next to each other experience each other within the limitations of relativity. The difference between these two units is mind-bogglingly small but is a finite amount. It takes great distance or great speed or great time for us limited beings to perceive these relative differences.

Lee Smolin, in his books *Time Reborn: From the Crisis in Physics to the Future of the Universe* and *The Singular Universe*, argues that while time is an intricate part of physics equations, it is wrong to turn that around to use these equations to create abstract theories that manipulate time. Time is real. It occurs just how we experience it. It evolves. Events are not predetermined and are dependent on our choices.

We have an "us" sense of time, which is completely discordant with the relativity model. We sense a feeling of now with a passing of the past and a movement toward the future. There is an arrow of time. Everyone and everything we know moves in it with us. We know this to be true as we can experience it with our own two eyes.

Every moment is special and particular. Once done, it is never to be seen again. Exactly at the moment I cough, someone on the other side of the world sneezes. These events occur simultaneously and independently. I could not possibly have access to the occurrence (instantaneous information) of that sneeze. It takes a certain amount of time for information to travel to me, but it was simultaneous just the same. Once completed, that moment is never to be experienced again. There is no chance for me to travel back to the moment of the sneeze.

Indeed, here it is important to note that Planck himself applied the concept of entropy not just to his work on energy and blackbody radiation, but also into the realm of time. As Brandon Brown notes, Planck took Clausius's theory that energy would never spontaneously flow from a cold body to a warm one—the beginnings of the concept of entropy—to a logical conclusion that ventured into the realm of time:

> The entropy of a system would never decrease with the passage of time. And because entropy would forever ratchet upward, Planck wrote in his thesis, "a return

> of the world to a previously occupied state is impossible." Or, the notion that you can never go home again applies to objects and machines—not just people. This was an absolute for Planck in his early career. He imagined a world that moved ever forward irreversibly, each event in each moment the exact product of the moments before it.[34]

But while Planck publicly held these views about the relationship between *entropy* and time, in his private correspondence, he also grasped the implications of the *quantum* on time—and even the then-new concept of space-time. In 1908, as part of a long, ongoing written correspondence with his closest friends, Planck wrote:

> Regarding Carl's question about my ideas regarding the elementary quantum, I must first confess that at present they are still pretty poor, but I would like to say the following... **the elementary quantum['s] full explanation cannot be done by considering a state, but only by considering a process. In other words: We are not concerned here with a theory of atomism in space, but with an atomism in time, via processes we usually think of as a continuous in time, that have, in fact, temporal discontinuities.** That the laws of ordinary mechanics and electrodynamics, which always presuppose temporal continuity, are inadequate here, may well be considered certain.[35]

The above emphasis is mine. Planck's published writing on entropy and his private correspondence on the quantum show the seeds of the argument that this chapter attempts to bring to fruition. At the smallest levels, the level of the quantum—that is, of Planck units—time moves forward in a single direction and, due to the second law of thermodynamics, each moment at that level unfolds unidirectional: forward. We cannot return to the exact state that just passed. By suggesting a "grittiness" to time and of *space-time*, just as Planck first grasped the grittiness of energy and Einstein first grasped the grittiness of light, Planck's private correspondence begins to suggest that "the fabric of the cosmos has a granularity, a grit, as opposed to being absolutely smooth and continuous."[36] And indeed, this grittiness of the fabric of the cosmos is what this book has been arguing for all along.

At the moment an atom of uranium-239 decays to neptunium-239 in the crust of the Earth, another atom of uranium thousands of light-years away also decays. These events are completely independent and unrelated but also unwitnessed and therefore not dependent on an observer. This all makes sense and fits reasonably well within our human intuition. For us, there is a particular moment and movement of time. It seems to be shared by those in our immediate surroundings if they are moving in a similar speed-frame, especially if that speed is nowhere near the speed of light.

There is an arrow in our experience. Playing cards do not get shuffled into order, eggs do not unscramble. We go from baby to child to adult to death, in that order. We can totally count on that order. It happens for us and for those around us in the same predictable pattern with the same predictable rate. Our watches all seem synchronized. Pediatricians have charts to make sure our kids are on schedule to follow the fiftieth-percentile rules of relativity. It seems each of us has only one arrow down this pathway of life. This pathway evolves based on chance and circumstances that we are presented with. This is further complicated by our personal choices we make in response to these circumstances. We only have one arrow each. This is our pathway, and it is unidirectional.

In 1905, there were some basic changes made to the framework. At our relatively slow pace and at our relatively short distances that fill our daily lives, it appears that we are experiencing the same arrow of time. It turns out that we each have a subtle difference in our space and time frames. On our human scale, this difference is so small as to be virtually impossible to measure. Even if we had instruments with this level of precision, it would make no appreciable difference in our experience.

On the other hand, at cosmic scales and astronomic speeds, the difference becomes readily evident. Suddenly, each of us has a position in time and space that is particular to us alone. Each of us has an equal right to claim ourselves as the reference point for all others, or maybe none of us has the right to make such a claim.

Not to imply that we are anything special in the grand scheme of things at the cosmic level, but we have evolved to a point of reasoning and decision making, self-awareness and consciousness. With this, we have also developed an enormous collective egomania about our importance. We still suspect that we hold some central value role in the universe. We suspect that our lives have some meaning outside of the little act we put on locally.

It would be silly to think that our lives were predetermined and that decision making is an illusion in a predestined series of events.

**Time is the stepwise regular unidirectional series that we personally experience, but is only truly valid at the Planck scale. Each Planck unit follows this orderly progression with each tick of the clock lasting only Planck time. There is no chance of variation at this scale and each unit has no choice but to move on to its next tick. There is no chance of mathematically manipulating time at this scale. Each Planck unit travels its own arrow.**

People claim ownership of a single arrow only because the time change from the tops of our heads to the bottoms of our toes at light speed is too brief for our measurements. As far as we can tell from our limited viewpoint, all of our Planck units are experiencing the same time frame. The time dilation between two neighboring Planck units is so infinitesimally small as to be essentially immeasurable and inconsequential. Even so, on at least a philosophical level, it must be recognized to exist.

While each Planck unit follows its own unvarying time course, relativity tells us that all other Planck units will "experience" our personal Planck environment in a way that depends on relative distance and speed of the units. Special relativity's time dilation and space contractions can be seen in the behavior of individual Planck units.

Even if a clock were devised that could measure time intervals at Planck time, it would be immediately invalid at doing so because the distance between the clock and the Planck unit would bring relativity into play. The only reason clocks work for our experience is that they are inaccurate enough to smooth over these discrepancies, and nothing in our lives requires that level of accuracy.

Each of us has a valid and spontaneous arrow of time. On even a smaller scale, every atom, every Planck unit that makes us up has its own arrow. Because of whatever consciousness is, we witness our collection of atoms as they travel this arrowed path together as one package. Each of us has our own "now." It is the only time we have personal access to. Others only have personal access to their now. For those of us standing close to each other, these personal present moments appear to coincide. In actuality, the time and space differences are just too small to measure.

At cosmic distances and near warp speeds, we would measure significant differences between participants. From my valid reference point in this scenario, my now will remain perfectly stable while you have become smaller and older. Within your equally valid reference point, your arrow has not deviated in the least, but I will have appeared to change to that very same degree.

You do not have to worry about waking up in the 1950s, kissing your mother, and making your brother's head disappear on a photograph. Your personal now is yours to experience and alter as you see fit. In theory, others could manipulate their now in reference to yours, and so your experience of them turns out differently than it could have, but once that path is chosen, it becomes part of the now for each of you.

## *Chapter 5*

# LIGHT

At the turn of the twentieth century, the existence of the aether, long held as a key component of many physical theories, was in serious doubt. Physicists scrambled to explain how theories dependent on the aether, namely those regarding the behavior of light, could still hold. The reason for this panic was 1887's Michelson-Morley experiment in Cleveland, Ohio, which seemed to disprove or at least strongly challenge the notion of aether's existence. As John Rigden notes in his excellent book *Einstein 1905: The Standard of Greatness*, the experiment "negated the accepted dogma of a static aether [and] implicitly raised questions about the aether concept itself."[37] Importantly, however, Rigden also notes that the aether was needed to propagate light, and there was no suitable alternative to the theory available: "In short, the experimental result had to be accepted, but the aether had to be retained," and as a result, "a series of ad hoc remedies were proposed to make the result of the Michelson-Morley experiment compatible with a static aether."[38]

Hendrik Lorentz, Dutch Nobel laureate and proponent of aether theory, was particularly perplexed and stirred by these developments. As a result, in 1892 Lorentz proposed one of the "ad hoc" solutions, albeit a solution would go on to help form the foundation of one of the most important scientific theories of all time: Einstein's special relativity. Lorentz's solution, "invented to patch over the problems engendered by the aether concept,"[39] suggested that under certain conditions, an object's length would contract. The Lorentz transformation equations, with their implicit suggestion of the relativity of a physical property as seemingly concrete as length, contributed to the growing discussion of relativity in the air. As Rigden notes, Henry Poincare would soon begin inquiring about the relativity of time. By 1900, some scientists were clearly thinking about space and time in entirely novel ways.

If Lorentz's equations are to be considered an ad hoc attempt to patch the holes in theory exposed by the Michelson-Morley experiment, then Einstein's June 1905 paper "On the Electrodynamics of Moving Bodies" is, as one of the great scientific achievements of all time, certainly a more thorough solution. In the paper, Einstein uses a thought experiment to consider the implications of his two new principles, the principle of relativity and the principle of the constancy of the speed of light, on the measurement of the length of steel beams under extreme conditions, and ends up with the same equations of Lorentz. However, Rigden notes an important distinction: "Lorentz's equations resulted from his ad hoc attempt to explain the 'failed' Michelson-Morley experiment; Einstein's were a direct consequence of his two principles and his conclusions about time."[40]

In an interesting twist, whereas Lorentz's equations were an attempt to save the aether, Einstein arrived at the same equations but, because of his two principles, found that his theory did not require the existence of the aether at all. In the paper, Einstein "asserts that, in terms of physics, the idea of absolute rest has no meaning. By this assertion, he rids physics of the aether. Einstein renders the aether, at absolute rest, 'superfluous.'"[41] It took several years, but Einstein's theory of special relativity was widely accepted by 1911, though there were still some holdouts among prominent physicists into the1920s and 1930s. But despite the resistance in its early years, Einstein's ideas on relativity, which were ultimately completed a decade later with his work on general relativity, laid the existence of the aether to theoretical rest.

Today, over a century later, a century in which the existence of the aether has not again been seriously considered, it is important to remember that the aether was dismissed, not because Einstein disproved its existence, but because his theory didn't need it. If Einstein's theory on relativity were a complete, airtight description of the universe, then it would be acceptable to completely dismiss the aether on these grounds. But even though the theory of relativity has stood the test of time—and it has—the existence of an aether compatible with this theory is still fathomable. Lorentz, for his part, was not ready to completely dismiss the aether, even while completely comprehending Einstein's theory, writing in 1913 that he "finds a certain satisfaction in the older interpretations, according to which the aether possesses at least some substantiality..."[42] Rigden uses this quotation to prove that Lorentz shows that even great physicists can underestimate the implications of revolutionary ideas, writing, "All of us, physicists and non-physicists alike, get comfortable with our common-sense ideas and hate to give them up" (103). But perhaps Lorentz was on to something. Perhaps both relativity and aether could exist in concert.

Three months before his paper on special relativity, Einstein had published another revolutionary paper, this one titled "On a Heuristic Point of View about the Creation and Conversion of Light." In this paper, the first of Einstein's major 1905 papers, he explained one of the great scientific mysteries of the time, the photoelectric effect, and argued that a light ray "consists of a finite number of energy quanta that are localized points in space, move without dividing, and be absorbed or generated only as a whole."[43] In short, Einstein argues for a particle understanding of light despite the fact that "[i]n 1905, the wave nature of light was an established, incontrovertible fact."[44]

As mentioned previously, the Thomas Young double-slit experiment from the 1800s provided definitive evidence of light behaving as a wave. Light from a single source shown through a single slit would be focused but also show some degree of scatter, which suggests a wave behavior of light. If two parallel slits were opened, however, an interference pattern would emerge. There is no possible explanation for this other than the wave behavior of the propagation of light. By 1905, Young's theory was orthodoxy. Prior to Young, Newton had suggested a corpuscular nature to light, but this had to be retooled in respect to Young's findings. An interference pattern is not what one would expect if light were corpuscular. Indeed, Newton's corpuscular notion of light was essentially its own form of orthodoxy that Young helped overturn. It was difficult to abandon the theories of one of the greatest minds ever, but the double-slit data were irrefutable.

In sum, we have seen several "orthodoxies" in our understanding of light replaced by more theoretically comprehensive and experimentally valid conceptions. The Newtonian corpuscular orthodoxy was challenged by Young's waves, which were in turn challenged by Einstein's return to particles. Rigden summarizes that "In March [1905], Einstein effectively made light both a wave and a particle" and that "Einstein called for 'a kind of fusion' of the wave and particle theories of light."[45] This duality is now a key concept in quantum mechanics and has in its own way become orthodoxy, though it has still not yet been thoroughly and satisfactorily explained. Perhaps this orthodoxy is ripe for its own overturning moment.

Wave theory would imply a continuum of energy levels. Any conceivable energy amount should be an allowable possibility. For any small energy value you give me, I could suggest a smaller value, and this degree of smallness would be limitless. Quantum theory shows that energy at its most basic level seems to behave as if it came in little steps rather than the infinite possibilities of waves. As Einstein's work on the photoelectric effect in 1905 shows, in order to explain some of the behavior of light, we would have to revisit the particulate model. As Einstein's proposed particulate nature of light does not reconcile with Young's, any new theory must reconcile the duality model of light.

An electromagnetic wave is not a physical movement of any particular element, just as sound waves are not a movement of air molecules any further than their vibrational motion. We certainly get the sensation of movement as we can time light from source to target with extreme precision. Electromagnetic waves, whose behavior is mathematically captured in Maxwell's equations, follow the same rules as all other waves. This would suggest that they are concentrations and rarefactions of some element. Believing and understanding this, moving from the wave-particle duality to a new, more complete understanding of the light wave, which still accounts for its apparent particle-like properties, will unleash all of the magic of the universe. The wave nature of light demonstrates the essential element that makes up the entirety of what exists and as it turns out, what exists out there as "nothingness"—the empty space in between stuff.

Light has a speed. It makes intuitive sense to us that it has a speed. We see a lamp over there with our eyes over here. It is pretty clear that something has traveled from one to the other for this event to happen. Even though the speed of light is too great for us to sense, we can still believe there was a time for light to travel that short distance.

Distance over time is the definition of speed, but what is traveling? A small, massless particle? A wave? Some magical hybrid of the two?

The concepts of the qualities of light can be broken down into its components. Electromagnetic phenomena move at the speed of light. Light has a signal strength we call intensity. Light has a wavelength. Light has an energy. Light has a pulse width.

Some of these components may, in some ways, be contradictory to each other, and some may seem counterintuitive to common sense. Nothing moves faster than the speed of light, and yet, as a natural wave, it is possible that nothing in particular is moving. For instance, light speed may be the universal speed limit, but even though speed is a cardinal parameter of motion, photons don't seem to follow the laws of classic motion mechanics.

To understand light in this new way that goes beyond, or rather unifies, the wave-particle duality, let's conduct a thought experiment to see how light could travel in a universe filled to every corner with what we have proposed to be the most basic, elementary units: Planck units.

At Planck length away from the light source, a "stream of photons" of light affect the first Planck-sized space. As the light leaves the surface of the flashlight, at only Planck length away from the emitter, in the very first Planck-sized space, that individual unit either receives enough energy to be activated or it doesn't. From a quantum standpoint, there is a certain threshold energy needed to deform this Planck unit. Below this energy level, there is no deformation (vibration); at or above this energy, our single index unit is fully deformed. This "all or nothing" nature is the very definition of quantum. The fact that we can so precisely define a single quantum of energy tells us that each unit deforms to a standard degree. Too little energy results in zero deformation, more than enough energy and 100 percent deformation.

The intensity of light, therefore, is not related to how much energy is pushed into a single unit, and yet the intensity (amplitude) is certainly somehow related to the amount of energy delivered. More energy, excess energy, will activate neighboring Planck units. The more units recruited at any given instant allows the transmission down the line in a more organized fashion, "in phase." The more units in a wave front that are in phase, the higher the amplitude of the wave. The total intensity of a light wave front is related to the number of units recruited in that wave front and the degree to which that wave front is synchronized.

A more precise wave front will correspond to a wave of higher amplitude and a sloppier wave front will disperse the energy packets over time, resulting in some interference and a lower-amplitude wave.

The degree to which the unit itself is deformed seems to be a constant. It is deformed or it isn't. It either vibrates or it doesn't. More energy can be transferred form the source either by activating more adjacent units or by increasing the amount of time a pulse is delivered. Once a wave front passes, an individual unit is prepared to receive its next impulse. Brightness may depend on the quality of the phase, but total energy delivered is also dependent on the duration, that is, number of waves.

The light pulsation must have some duration. It must activate the initial unit over and over again in order for a detectable signal to pass down line. The number of activations or vibrations per time is the definition of frequency. Any Planck unit can be activated as frequently as the light source dictates up to the limit of Planck time. The higher the frequency, the higher number of wave fronts established and the shorter the distance between wave fronts, that is, shorter wavelength. Frequency is the inverse of wavelength.

An energy transfer may be limited at the individual Planck unit for a single vibratory pass as a single quantum of energy, but total impulse energy is dependent on how many of these individual vibration waves pass. If an impulse has enough energy to cause this vibration, one quantum, the total amount of energy transferred is proportional to its frequency.

We see that light does not follow the rules of Newtonian mechanics. A car moving at sixty miles per hour will appear to be moving sixty miles per hour to a bystander on the side of the road. It will appear to be relatively stationary to the car in the next lane going at exactly the same speed. It will seem to be going 120 miles per hour to the oncoming traffic.

A bullet shot from a 30-06 travels at 2,820 feet per second at the muzzle. Shot from a car moving sixty miles per hour, add an additional eighty-eight feet per second that the bullet would appear to be traveling to the stationary observer and even an additional eighty-eight feet per second to the frightened oncoming sixty-miles–per-hour traffic. In classic mechanics, all of these velocity vectors can be summed.

Light doesn't have a muzzle velocity; it only has its one and only one velocity, which is "the speed of light." Shoot a bullet from a gun while traveling in a car, and the bullet speed will depend on the car's velocity. Shoot a flashlight out the front or back window, and no matter what the vehicle speed, the light will be measured at 670 million miles per hour. If the car were moving at one hundred million miles an hour, it still wouldn't change the measured speed of light as measured from that car. A bystander measuring the light would also measure *c*. If a bystander with a flashlight were shining it at the front or back windows of the car coming or going, the car's passengers would measure the same speed of light. This goes against our intuition and Newtonian physics.

Light, in a vacuum, travels at the speed of light. A light from a stationary source travels at 186,000 miles per second. A light from a moving source still only travels at the speed of light. If our solar system is moving in the spinning Milky Way at 300 km/s, a light emitted from a planet in that galaxy still will travel at *c*. It doesn't get the boost from the Milky Way that our 30-06 bullet gets from our moving car. Light, whether sent from a stationary source or one moving toward us at nearly the speed of light, will still be measured at *c*. This is the universal speed limit, and it does have a strict governor.

The reason for this is that light is not dependent on the source at all. As soon as it leaves its source, it starts its travel through its Planck units. The velocity is totally dependent on how it moves through these units. It does so in a universally uniform and predictable fashion. In doing so, it demonstrates the uniformity of these Planck units. As soon as light leaves its source it is independent of that source. It travels independently until it reaches its target, which is your eyeball or some detector. In the instant that the one Planck unit's worth of light strikes the detector, it is traveling at the pace of light.

As a star moves away from us, the light we detect is moving at *c*. As the star recedes, it sends off its light into the Planck matrix so that it is transmitted one Planck to the next at *c*. We measure one of light's properties to determine the speed of the star relative to us. This measurement is our evidence of an expanding universe. It is the Doppler shift of frequency as measured by an observer. It is the red shift of stars moving away from us.

Light is transmitted in the same way that nature transmits all waves. The energy leaves its source, vibrating the first Planck unit, which is in direct contact with its surrounding units. This index unit transmits its energy to these surrounding units and does so at the speed of light. As with all waves, no unit moves more than its vibrational, back-and-forth motion. No little, massless particle is actually traveling from flashlight to eyeball.

Max Planck's wristwatch was especially precise. Planck time is the amount of time for light to travel Planck length.

Planck time is time's absolute zero. No time is shorter than the snapshot of Planck time. The tick of Planck's watch.

As light vibrations are transmitted Planck unit to Planck unit, they do so at the speed of light. In this simplified model, they move down the line from the source, through space, and then possibly to a point of detection. As light transitions from the last Planck unit in "ordinary space" to the first Planck of the detector, it does so at the speed of light. Again, it does this within our new definition of instantaneous as it occurs at Planck's absolute zero time.

An electromagnetic event is the summation of vibrations of all involved Planck units. The Planck unit is the shortest possible distance, so the vibration doesn't actually travel "through" it per se. Through it would suggest the vibration moved from one end to the other, but there is no possibility of movement when the Planck unit is already the shortest possible distance. In Planck time, the vibration of an electromagnetic event has not traveled through a Planck unit, it existed in a Planck unit. In that brief Planck time, all electromagnetic activity exists in their respective Planck units. In the next tick of the watch, every electromagnetic vibration in the universe has moved exactly one Planck unit. A light vibration moving east to west will have moved one and only one unit. A light vibration moving west to east will have moved no more and no less than one Planck unit.

The specific quantum size of Planck units is why light has its intrinsic speed, *c*, rather than a variety of speeds. It is obligated from one unit to the next (Planck distance) in an exact amount of time (Planck time). Color depends on the number of activations of Planck units per unit time, but all colors still travel at the same speed, suggesting that travel "through" each unit occurs at only one rate, and this rate is the natural harmonic of all Planck units.

It seems quite perplexing that the speed of light should be an absolute constant and not depend on the additional velocity vector of its source or the velocity vector of its observer. This does indeed violate classical mechanics.

In space-time, an electromagnetic transmission is purely though space so that the time dimension is absolute zero (Planck time). In the next Planck time, our light vibration has traveled one more unit but the detector is also traveling on its own arbitrary way. In that same instant that the vibration got translated to the same space as the detector, the detector is frozen in time, our absolute zero time, as is all of the universe. Light and detector share that Planck space as a coincidence for only that briefest of all possible times. No information can possibly pass faster than this universal speed limit. So even if the detector had been traveling through space at 0.5 times the speed of light, that photo-finish snapshot encodes that moment at the extremes of space and time as the speed of light.

The spread of this impulse is not linear. It spreads out as it propagates. It is a wave front. Every Planck unit in proximity to the first will then be affected. This idea was postulated by Christian Huygens, who argued in his 1690 *Treatise on Light* that "one may conceive Light to spread successively, by spherical waves" and that each source of light produces a "prodigious quantity of waves."[46] Every point on a wave front can be considered a secondary source of spherical wavelets. As Huygens describes it:

> [I]n the emanation of these waves ... each particle of matter in which a wave spreads, ought not to communicate its motion only to the next particle which is in the straight line drawn from the luminous point, but that it also imparts some of it necessarily to all the others which touch it and which oppose themselves to its movement. So it arises that around each particle there is made a wave of which that particle is the centre.[47]

A more focused light source suggests that the concentration of energy will be more in the direction of the impulse. Even in a focused impulse, the more forward units will receive more energy transfer, but every surrounding unit is affected to some degree. There is always scatter. A laser will send more of its energy downline than a spotlight. A spotlight will send more of its energy downline than a floodlight. Even the most focused energy event, however, will scatter to some degree. Some energy will even be bounced backward to a minuscule degree.

This does not jive completely with the quantum event of Planck unit deformation. The energy of a single deformation is a quantum of energy and is all or nothing. Any scatter would suggest an energy leak in the system, and so immediately at the second Planck unit, there would be insufficient energy to propagate the deformation.

At the individual event level, the energy transfer is a yes or a no, 100 percent or 0 percent. The degree of energy transfer, focus, and scatter are a function of the quality of the wave front (how pure is the pulse phase of the wave front) and the frequency of wave fronts.

If the energy of a wave front can be shown as the area under the curve of a wave, then a more spiked wave will concentrate its energy mostly under the peak of the wave. A more slurred, gentle wave may still have the same total amount of energy (area under the curve), but at any given instant, the concentration of energy is less. The intensity of an electric event is proportional to how high this peak is, that is, how many units are in phase? This is the definition of amplitude.

A photon flying through space should follow classic mechanic rules. Vectors should be summed. If, however, a "stationary" Planck unit is responsible for transmitting that information, the rules of motion might not apply. As mentioned, Lorentz was a proponent in general of Einstein's theory but saw that these results could still be achieved in the setting of an aether framework. The quantum unit of quantum physics is the Planck unit. These units are the corpuscles in light's motion. The units are not in motion but are the medium through which light moves. This is the real unification between the wave and corpuscular theories of light: light moves as a wave *through* the corpuscles.

Special relativity gives us insight into how light velocity is measured by observers in nonaccelerated motion, but everyone measures the speed of light exactly at 670 million miles per hour. That might be the case if the speed of light is dependent on the medium through which it travels, rather than on its source.

It is funny that although most scientists would be uncomfortable considering such a substrate as aether, most are quite comfortable replacing it with another, similar all-pervasive goo. The Higgs field, for instance, is that same goo by a different name. Even Einstein eventually tried to replace it with his cosmologic constant, which has a more math/science ring than the magical substance of aether. The main benefit of the terminology shift is that it feels less like alchemy and more like modern physics.

The stream of light energy passes through these Planck units at a certain frequency. We detect certain frequencies as a specific color. As the source recedes away from us shining its yellow light, the number of cycles per unit time doesn't change, but with each ticking moment each light unit will start at further and further Planck units away from our retinas. The result is that although the light will travel through these units at the speed of light, the rate of the cycles per unit time will reach us more slowly. This will give the appearance a shift to a lower frequency. For visible light, this would be a shift toward the red part of the spectrum.

In a collider, two particles are shot in opposite directions. They meet each other downstream at a predetermined focal point. If one particle were fitted with a speed detector it should measure a relative speed of the approaching particle at nearly 2 *c*. This is an impossibility and a dilemma. The information of the speed and even the existence of the approaching particle cannot pass through the Planck units faster than *c* to hit the speed gun detector. As this information reaches each Planck unit, the shortest possible length, it does so in Planck time, the shortest possible time. These two values, Planck distance and Planck time, define *c*. There is no space and time that can breach these barriers. It is the definition of instantaneous. It is a true snapshot. Look at a photograph of a moving car, and it appears stationary. One cannot look at the photo and tell you the car's velocity.

Electromagnetic events do occur in wave form. We sense a movement in light's behavior, and our intuition wants to attribute this to a particle. We had to invent a fake particle, the photon. A photon is not a moving particle. It is an instantaneous snapshot of one Planck unit as an electromagnetic impulse passes through. The impulse activates the Planck unit in some way that we might consider vibration. Furthermore, vibration is just our word that fits our macroexperience but at that small scale, who knows if that is an appropriate term or even a reasonable model?

As each unit is essentially identical to the next, impossible to distinguish from the next, it might appear to the bystander that the units are actually moving. The energy is transmitted amazingly fast—at the speed of light, in fact. It gives the impression of motion.

The duality of electromagnetic phenomena is thus evident. It is a wave in that the source emits it with a certain frequency and amplitude. It seems to spread out like a typical wave. That first Planck unit can be vibrated a thousand times a second or a million times a second, but that unit is affected individually one vibration at a time. That unit couldn't care less. It just passes on its vibration for Planck length, at the speed of light, never giving it another thought and never moving more than its own internal vibration. The image of that Planck unit in action transmitting energy down the line allows us to imagine that unit in its snapshot form as being the moving unit: The photon.

**The Planck unit is the photon.** It is that massless speck that makes us think of a particle. It transmits its vibration as all natural systems do, by passing it on to its neighbor in a series of compaction/rarefaction events known as a wave.

So let's build a model of an electromagnetic event. We use the term "light ray," but light disperses in an ever-expanding wave front. I will use this simple but incorrect image of a ray to start building our model. Then we can add the many layers of complexity to this vibration once a simple model is established. In the simplest model, let's consider only a single ray of Planck units transferring energy without any dispersion or interference.

We start this process with a large "agar block" of Planck units in the quiescent, undisturbed state. As we turn on our flashlight, in one Planck tick, the first Planck unit is ever so slightly distorted. This distortion is transmitted as a vibration to the next neighboring unit. Ten ticks later, and this impulse has moved ten units down line. This simplified model of the passage of vibratory information down line is akin to our model of a photon travelling as a single ray. This image is far from the full behavior of this energy transmission.

The next layer of complexity has to do with the fact that no beam of light is Planck length in diameter. In the idealized, perfect model above, a single Planck unit is disturbed and this wave of disturbance is passed down the line unit to unit. While from an educational standpoint it may be beneficial to consider a light event at the single quantum level, that is, not the actual essence of waves. Emitted energy leaves a source as a sizable chunk. In nature, an energy source event does not occur at the single quantum level. Instead of our simplified model where a quantum sequentially affects (vibrates) Planck unit after Planck unit from source to detector, a swath of units initiate and transmit this energy. In doing this, there is a compaction and rarefaction of a swath of Planck units. Much like a single water molecule moving up and down would not make much of a splash, a single ray of Planck units would not make much of an impulse. There is a thickness to the wave form transmission of light. Indeed, Huygens recognized this concept over three hundred years ago when he explained why we're able to see the light of such distant bodies as stars: "...an infinitude of waves, though they have emanated from different parts of the body, unite together in such a way that they sensibly compose one single wave only, which, consequently, ought to have enough force to make itself felt."[48] What Planck describes as this single, "master" wave, we can apply to Planck units as we consider the aforementioned swath behavior.

We are still far from the truth. So far, we have multiple rays moving out in all directions, creating a wave front, but each ray is responsible for passing its own parcel of vibratory information down the line.

The next layer of complexity is also a supportive of the Heisenberg Uncertainty Principle for electromagnetic phenomena. The deformation energy of the first Planck units transferring 100 percent down the line to the next units. In actuality, a bulk of the energy is passed forward, but a little is transferred to every unit in direct contact. The degree of this scatter would depend in large part on the focus of the light source, how parallel the emitted waves are produced. No matter how focused, some of the deforming energy would be transmitted to nonlinear units.

Not only is Heisenberg satisfied by this process, but this is also why entropy is an obligatory law. As these vibrations are transmitted downstream, they become diffused throughout the general path rather than that streamline sinus wave we picture. Not only is the information diffused but also further confused with the input of information from the nearby units trying to vibrate their information about our flashlight and furthermore from other flashlights that happen to be intersecting our light transmission.

Let's now blend these two models together. A chunk of energy initiates an impulse involving a bunch of Planck units. Each unit sends its information down line through this vibratory event in a wave front scattered pattern. With this scatter, each unit layer would have this energy more and more dispersed so before long a single unit's energy would diminish toward zero. This would inevitably and pretty quickly lead to not enough energy to vibrate a Planck unit and therefore the end of a light transmission. Since light emission occurs as a chunk of energy and not just a single quantum, as units in the original ray path would be losing energy they would also be receiving scatter from every nearby unit. It is the summation of this scatter which accounts for the light that strikes our retina. If we had a Planck-size Polaroid, we could see the snapshot image of these energies zipping in various directions but no specific ray holds any specific energy. This fuzziness of photo transmission will not allow us to define the Planck unit and direction for a single packet of energy.

In terms of Heisenberg, no specific unit owns the totality of an impulse and so from a photon model, one could not define the location of that specific particle. Still, that is only a partial account of the uncertainty principle. The other part of uncertainty has to do with the speed of transmission of information. This occurs at the speed of light. Unless you, as an observer, lived inside a specific Planck unit, there is no way for you to witness the particular extent of its involvement in any event, because by the time the information has made it to you, Planck's wristwatch has ticked that many times, and the complex event is utterly changed.

Even with all of these layers of complexity, there is still more to the light event. It is absolutely vital that light emission initiates as a chunk rather than a point and travels as a swath rather than a ray, or as the impulse wave spreads out, it will not sustain enough energy to vibrate units down the line for very long. This suggests that a wave-front impulse has a thickness or a measurable area. I would suggest that it also has a measurable "length."

All waves behave the same. Drop a pebble in a still pond and what you see is not a single crest but a series of crests and troughs. Flick a tuning fork and what you hear is not the single deflection of the tines but a persistence of sound. Move a ball on a string from its resting point, and gravity will naturally pull it back toward the center of Earth, but even though that is its tendency, we would be shocked if this pendulum reached its center and just stopped. We know it will pass beyond center before starting a new and repetitive cycle.

Light is a wave, and all waves behave the same. Deform a Planck unit to initiate a vibratory impulse, and it will naturally spring back toward normal. It would be unnatural and unwavelike to expect the deformed unit to spring back toward its resting condition and not overshoot. Every electromagnetic event must have some persistence. Even if we somehow developed equipment that could deliver an impulse lasting only Planck time, the natural ripple effect of waves would result in persistence.

Meanwhile, at ninety degrees, every time a photon wave has passed, a magnetic event has also occurred. Magnet and magic were probably put pretty close in the dictionary on purpose. The idea that magnetic fields can have an effect on something else at a distance, without touching it, is truly magical. Fortunately, we no longer need sorcery to explain the existence or transmission of a magnetic field.

Again, waves are nature's way of transmitting information. All waves, at a fundamental level, behave the same. I will use the ocean wave analogy again to show where magnetism lives in an electromagnetic wave event.

When we imagine an ocean wave we picture the horizontal movement of a crest from offshore moving toward the beach. This is akin to the transmission of an electric impulse. Remember, though, that in an ocean wave, no specific water molecules move to any significant degree toward the shore. It is only the cyclic compaction and rarefaction of water molecules that make up the peaks and crests. Only the information is transmitted forward.

A device could be made with a float that would move up and down with the ocean's waves that converts the up and

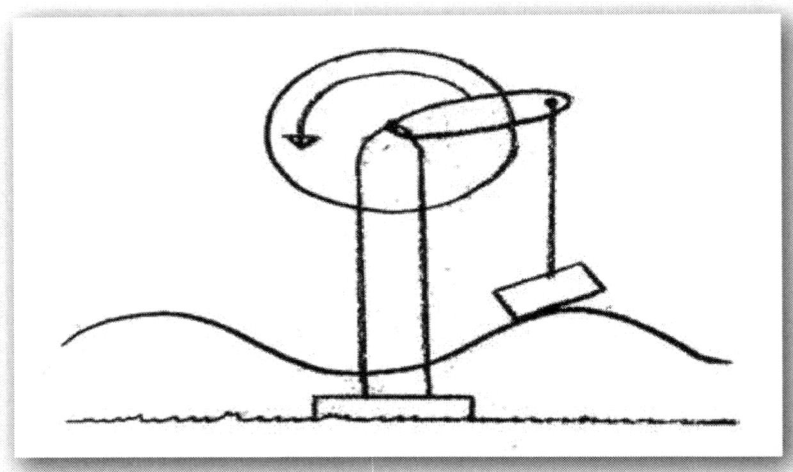

*Figure 3: A turbine that harvests energy from a floating bob's vertical movement with each passing wave.*

down motion to turbine energy. Instead of picturing the energy transmission of a wave as moving horizontally toward shore, we can now clearly see that there is a harvestable energy that occurs in a perpendicular direction.

Since we know that all waves are natural phenomena and behave with similar properties, an electric wave can be compared to ocean waves. Light has a perpendicular equivalent to the ocean wave, and this is the magnetic portion of the electromagnetic pulse.

I will again use the word "vibration" as if we understood the nature of this energy transmission, which we don't. So as not to go through too much verbal gymnastics, vibration is the model I will stick with. In this vibratory process of the electric impulse, there is a moment of positive and negative deflection, compaction and rarefaction. While each individual unit is experiencing this transformation we know that the wave occurs as a cumulative gathering of a swath of units. Although there would be some general inefficiency in impulse transmission, some sloppiness to the wave, there would be a general tendency in nearby units to participate in the trough or peak of this wave. This results in a local and perpendicular alternating

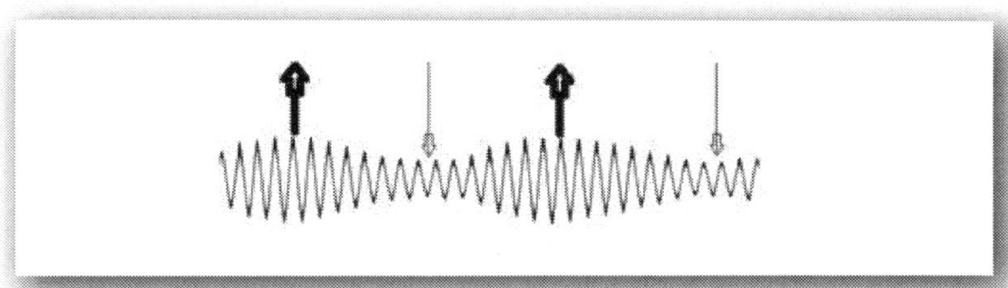

*Figure 4: The perpendicular push and pull of a wave.*

push and pull. The cumulative vector of these displacements especially in a continuous current is measurable as a magnetic field.

The magnetic field is transmitted perpendicularly from the line of electrical force from Planck unit to Planck unit so it is in this fashion that it seems to act at a distance from the current source. There is no magic necessary. Just as an electric impulse uses Planck units to appear to move forward, the magnetic impulse uses Planck units to move laterally, at the speed of light by the way.

Once we accept this model, we can look to an even further-reaching implication, which I will preview here but explore more deeply in subsequent chapters. Magnetism is most easily visualized in an electrical current system, but electrostatic models offer some important insights as well. Charles-Augustin de Coulomb's work on electrostatics, based on earlier works by Priestley and Cavendish, showed that the electric force between two point charges is directly proportional to the scalar product of the charges and inversely proportional to the square of the distance between them. This may seem a fairly innocuous and acceptable statement, but it's remarkable in its similarity to Newton's inverse square law of gravity. This may be mere coincidence or may suggest a link. Is it possible that magnetism is to electrical transmission as gravity is to mass?

## *Chapter 6*

# EXTRA DIMENSIONS

Man has always had an interest in classifying and quantifying the world around him. Geometry has been used by man since we first needed to know how to build our shelter or shoot an arrow at an enemy. Before any rules, proofs, or theorems were put forth, measurements were made in an instinctive way. Lengths, areas, and volumes followed predictable patterns. Eventually, in both Eastern and Western traditions, geometry began to be systematized, and continues to evolve, to some degree, into the present. The study of shapes and measures that began centuries ago has, in the past hundred years or so, taken a turn toward the quantum. The genealogy of geometric theories is most certainly prehistoric, but it is always right to pay homage to the lineage we know.

Thales lived in Miletus, now part of Turkey, in the sixth century BC. He studied abroad, notably in Egypt, where he adapted the mathematics of building pyramids into formulae for measuring triangles. He has been dubbed the "father of science" and the first sage of the "Seven Sages" of Greece, whose lives spanned from the midseventh to midsixth centuries BC. Thoughts attributed to Thales span the disciplines of geometry, physics of matter, politics, and business. They were the beginnings of the scientific revolution that started to look at reasoned explanations for natural phenomena rather than attributing them to the mythological stories.

It is said that as a child, Pythagoras met and was significantly influenced by Thales. Pythagoras of Samos was a Greek mathematician and philosopher from the fifth century BC. Like Thales, Pythagoras is believed to have learned a great deal of the principles of geometry during his travels abroad, especially to Egypt. Pythagoras's teachings were the basis of a school of thought that had many very strict beliefs. The foundation for these beliefs was that all of the world around could be defined by numbers. Notably, his school is likely responsible for the Pythagorean Theorem that defines the angles of right triangles.

Euclid was a mathematician from Alexandria around 300 BC. His text, the *Elements*, is considered the foundational geometry text. It lists the axioms and defines the theorems through proofs that made sense of geometry. The rules of Euclidean geometry continue to work in the world we experience each day. Euclidean geometry makes sense in our three-dimensional world. This was sufficient from then until the nineteenth century.

Janos Bolyai was a Hungarian mathematician. His father was a well-known mathematician of that time and instructed him in complex math, including calculus, which he reportedly mastered by the age of thirteen. Bolyai was apparently consumed with the fifth postulate of Euclid's parallel lines. By 1823, he had prepared a formal work on the geometry of parallel lines on curved surfaces that violate Euclid's postulate.

Traditional Euclidean geometry was fairly straightforward—or at least very well established. In order to examine curved surfaces in a mathematical sense, there needed to be some way to convert the new concepts under consideration to the established rules, making a three-dimensional object follow the rules of a flat object. This is what we do when we make a flat map of the spherical Earth. When done through equations, the flattening conversion terms are called tensors.

Bolyai's father was a lifelong friend of Carl Friedrich Gauss. A copy of Janos's unpublished findings was reportedly sent to Gauss, who responded that he, too, pondered the ideas of measurements of parallel lines on curved surfaces, and his findings were similar. This was a crushing event for Bolyai, who felt that he had formally discovered non-Euclidean geometry. As a result of this lack of enthusiasm, his work was only ever published as an appendix to one of his father's texts in 1832.

Around the same time, Nikolai Lobachevsky, student of another friend of Gauss, reportedly "independently" developed hyperbolic geometry, which is also a non-Euclidean geometry. He published his work in 1829, three years before Bolyai.

There was some controversy about the timing of these works and to whom to assign credit. Whatever the truth may be, hyperbolic geometry was referred to as Bolyai-Lobachevskian geometry. In any case, the turn of the nineteenth century was witness to geometry's first major revolution in thousands of years. This revolution set off a chain reaction of innovation in geometry that has continued to the present day.

A student of Gauss named Riemann was given the task of exploring the theoretical possibilities of extra dimensions. This work, published in 1868 (two years after his death) resulted in yet another level of new geometry. Riemannian geometry would eventually be the basis through which Einstein developed general relativity and curved space-time.

Using the basic field equations of Einstein's general relativity, Theodor Kaluza, from the University of Königsburg, in some very imaginative work, added one more possibility in the tensor position reserved for the three spatial dimensions and one time dimension that we experience in everyday life. It is difficult to understand why this possibility might have come to him, after adding this new fifth dimension to Einstein's equations, Kaluza found that new equations could be simplified to resemble those of James Clerk Maxwell's original electromagnetic theory. This tweak to the calculations yielded two astounding findings: first, that some aspects of matter could be compared mathematically to electromagnetic events, and second, that extra dimensions beyond the ones with which we're familiar are theoretically and mathematically possible. In addition, and particularly interesting for the present discussion of our proposed Planck unit model, Kaluza's calculations "indicated that the additional circular dimension might be as small as the Planck length."[49] So in the early twentieth century we had both Planck's quanta and other, independent calculations that proposed the theoretical possibility of additional dimensions existing on the Planck unit scale, though, as we will see, these ideas would not begin to be combined for decades, and have still not been distilled into their most logical conclusion.

That matter and electromagnetic phenomena might be related in some deep essential way was amazing in itself and had never even been considered, but the concept and possibility of further dimensions opens up a great new paradigm. Unfortunately, the initial math did not initially hold up to further scrutiny as attempts were made to include particles. This "hole" in the theory allowed the concept of extra dimensions to get lost for a while. The likelihood, however, that the similarity of Maxwell's electromagnetic equations and Kaluza's adaptation of Einstein's equations represented a mere coincidence was unlikely, and while it remained dormant for decades, Kaluza's idea that gravity and electromagnetic theories could be combined—in addition to the potentially even richer possible existence of extra dimensions—again came to life in the 1970s.

Riemannian geometry had only theoretical applications until the advent of string theory. In the midst of the amazing mathematical equations of string theory came also some very odd results. Initial attempts at string equations left unacceptable infinities and negative probabilities. This is an absolute dead end for any math-based theory and suggests that a theory is moving in a wrong direction. In addition, there was the further challenge that while extra dimensions may offer results that fulfill mathematical logic, they seem to have no place in our common experience.

In particular, one subfield of string theory tries to describe how strings exist in space. How they are "stretched" to make shape and volume. In these attempts, equations were manipulated and massaged with amazing, almost unbelievable results. It is here in the mathematical term describing the coordinates of spatial dimensions that more than three numbers fit. Mathematically, in order to allow the calculations to render acceptable results, they had to manually insert some number, a degree of freedom, a fudge factor. In string calculation it's not just that more than three dimensions could exist, but these extra dimensions were *necessary* to make those dead end infinities go away. So as an unexpected and exciting side event from the mathematics of string theory, the anomaly of extra spatial dimensions has a renewed and firmly established foothold. While Kaluza and Klein showed that extra dimensions were mathematically possible, string theorists began to argue that they were necessary. These extra dimensions have been the subject of exhaustive research in the intervening years, but let's now turn to how our Planck units could fit into this extradimensional universe.

These very smallest spaces (Planck units), the very limit between nothing and something, have a variety of potential manifestations. They can exist, for instance, in their most quiescent, undisturbed state. In this state, one could even imagine them assigned a position in the universe using our standard Cartesian three dimensions creating a structural framework. These undisturbed units would not use or exist in any of the extra dimensions. They are essentially a point. If a thing could exist in one dimension it would be this singular essential point.

On the other extreme, Planck units can be tied up in a matter-form. In this familiar form, our three dimensions define where the entire collection of units as a whole resides in the universe, a chunk of charcoal or a table. The individual Planck units are hidden from our perception, swept up into the complex configurations of these extra space-dimensions. It is these complex configurations that make matter. They are predicted by string theorists as complex manifold forms and thought possibly to reside at the very smallest level.

If it is the case that matter is a complex of Planck units hidden in extra dimensions, the three classic dimensions might only be important for our perception. The matter that we so strongly believe in may really be a resultant image of the folded units. What we see as a table may be only a backdrop for the real action, the real players, the swath of fabric enveloped into the extradimensional folds. The folds of Planck units are real but exist on scales far too small for us to fully perceive, and what they create could be something suggestive of a hologram that happens to be their only essence available to our senses.

It is almost certain that these extra dimensions are more than a mere mathematical anomaly or part of a fun theoretical exercise. Again, Kaluza's work shows that this is much more than an ingredient for science fiction, and string theory shows that rather than a glitch in the equation, the extra dimensions are an absolute necessity.

If these extra dimensions represent an actual physical space, where can we find them? On one hand it might be helpful to consider these extra dimensions in the way we consider the traditional three spatial coordinates. On the other, it goes against our logic and common experience to consider these spaces in a traditional manner. If these extra dimensions could be considered apart from the other three, we might gain some insight into their importance.

A single Planck unit will not allow for Riemannian geometry. A single unit will not fold into complex manifolds. If we allow for these Planck units to act collectively, though, we could imagine how a strip of them could result in various useful patterns not unlike the patterns found in Riemannian geometry or in the Calabi-Yau shapes. A specific pattern that lends itself nicely to the mathematical manipulation of special geometry is the string, a linear array of Planck units. Not only will it permit the required math, it dramatically advances it.

It is even more likely that rather than our notion of a string—a linear array—these manifolds are complex gatherings of a swath of three-dimensional fabric of Planck units. Sometime in the remote past, some enormous energy twisted trillions of units from all directions into this collection that we recognize as matter. Think of the energy expenditure to take trillions of Planck units to create one lowly quark and then multiply that by all the matter in the universe.

It is truly difficult to wrap our heads around the concept of more than three spatial dimensions. The use of the phrase dimension has its charm but it may be that dimension is not the best descriptor. But as the word is firmly established in the String theory literature, it is likely here to stay. Furthermore, the extra dimensions' existence was found to occur in the dimension tensor position of Einstein's original field equations. In a way, it is unfortunate that they have been linked to the traditional 3 + 1 Cartesian dimensions, rather than having a more descriptive identity of their own.

From a mathematical standpoint, according to Kaluza-Klein, there could be an infinite number of dimensions, but each one would require a much higher level of complexity. This complexity becomes unwieldy after these nine calculable dimensions—or possibly a tenth, plus time. This is a mathematical construct and is not to say that all nine exist or ever have existed. But from a mathematical standpoint at least, nine gets rid of the mathematical inconsistencies of negative probabilities and infinite (nonsense) values.

It might be better to assign dimensions five through nine their own definitive domain. They could represent vibration, spin, charge, and knottiness, among other possibilities. Once we have better means to explore these spaces we can determine more descriptive terminology. For now, dimensions are just anthropomorphic terms to help deal with our mental model. In the ethos of the string world, it would be sacrilegious to describe them in any other terms than dimension at this point. Dimension sounds so much more palatable than fudge factor. The curled up Planck units with their specific vibrations and spins represent the manifest actions of strings in their complex matter form.

Applying the concept of Planck units to the string discussion allows us to see the theory in a new light. At Planck length there are zero dimensions. In a ray of Planck units, we would predict a single dimension. A plane of Planck units yields a two-dimensional existence. In our familiar world, large Planck collections (matter) are visible in the right-left, back-forth, and up down directions.

Our familiar Cartesian dimensions in empty space are made up of these Planck units in their stable, undisturbed, featureless state. In order to transmit light, a wave of units must transfer energy (vibrate) one to the next. In that vast array of Planck units that make up the universe, the ones that we are presently interested in as a "ray of light" traversing a direct path from flashlight to eyeball represent a string. The string in this case is not a fixed entity but whatever grouping of units happens to be of interest at that moment. Using the string model, collections of Planck units become strings. Strings are momentary arrangements of Planck units. They exist and exhibit their qualities in ever increasing complexity to accomplish the task of forming mass and charge. They do this by utilizing ever increasingly complex dimensions.

It is thought that strings display their function by their harmonic vibrations. A quark would need to vibrate continuously to maintain its existence in this schema. This implies that strings must have been imparted this vibrational energy on the first day of creation. This vibration energy would have had to vibrate that string ever since without being used up, which, of course, violates the second law of thermodynamics and therefore is not allowable.

Vibrations and energy do not behave that way. Vibrations are energy dependent, energy consuming, and subject to entropy. There would need to be a continuous albeit small source of energy to keep the vibration active. To maintain the specific character of that particle the string vibration would have to remain forever active, no change in pitch or volume. While we might prefer a nonentropy vibrational system so as to explain the locked up stable energy of a particle, the gods in charge of thermodynamics forbid it. We need to reconcile the patterns of string vibrations as they appear to us as the masses and charges of the elementary particles. One possibility is that the geometric size and shape of the extra dimensions dictates the fundamental properties of the universe.

String theory is a unifying structure linking the essential units of the universe to the lights and particles that make up that universe. At one moment, a string is a collection of Planck units that happen to be participating in whatever happens to be interesting to us at the time, but at the same time they are not a permanently part of that structure, the participation of those units is transient, even if the particle is not.

We maintain the idea of the energy of vibration as giving the string some sort of structure but, as mentioned, a vibration would succumb to dispersion and entropy. Matter would not be sustainable. Part of the problem is due to the term "vibration" and our assumption that it represents what we experience on the macroscopic scale. Vibrations and windings represent the uptake and passage through ordinary space that Planck units experience as they exist ever so briefly in the complex shape of extradimensional space.

Whether string theory is a valid model or not, Planck units likely participate in the formation of particles and it may have something to do with the particular character of these complex shapes. There is a particular field of mathematics called manifold theory. It is a branch of algebraic geometry and has a long lineage with many participants, most of whom labored away with little recognition.

Based on Riemannian algebra, complex curved structures can be created mathematically. These structures can take many forms but some of the closed structures represent manifolds. These manifolds can have smooth features or faceted features and are called holomorphic.

Erich Kahler, a German mathematician, formed an algebraic, differential metric to help define these manifolds. His work was based on group theory of Elie Cartan and non-Euclidean geometry of Riemann. This work was complex and theoretical with little or no utilization in everyday life but was a strong clarification of Manifold theory.

Eugenio Calabi, an Italian mathematician, further advanced Kahler's equations to create specific conditions for certain types of manifolds, but these complex equations did not always yield acceptable results. This work sat on the shelf for some time until in the mid-1980s when Shing-Tung Yau worked out some of the details, through which Calabi's conditions provided sound structure. Through these formulations, he discovered shapes that exist in multiple dimensions. It was the six-dimensional structures that finally resulted in reliable findings without any contradictions. These were later called Calabi-Yau manifolds. It is an offshoot of this same math that had led Einstein to the tensor model of his field equations that are known now as the theory of general relativity. Even so, manifold theory was still purely empirical and occurred well before there was a string theory in need of it.

At first, there appeared to be only a few specific manifolds but as it turns out there are tens of thousands of examples of Calabi-Yau shapes that could fit the string theory model for extra dimensionality. This is a shame because matter is not made up of tens of thousands of varieties of quarks but more likely only a few. So the question arises: are these manifolds actually a general model of the true existence of some manifolds, or are there only a select few true Calabi-Yau manifolds that exist in nature, and these represent all of what matter is? If so, which of these Calabi-Yau manifolds are the chosen few and how did nature decide? How do we discover them? Is a quark made up of one manifold?

The strings we know in our daily lives vibrate and so in the terminology of String Theory, vibration is the name of the mathematical position, which describes that extradimensional behavior. The vibration they refer to is a continuous eternal intrinsic vibration that give a string one character versus another. It is not quite the transient vibration of a complex of information passing through a swath of Planck fabric. It is supposed that the eternal vibration is an expression of the enormous energy required to wind a long string into a very small space of a manifold.

Another common term to express a complex concept in manifold theory is "hole." Calabi-Yau shapes are said to contain holes, which actually can exist in a variety of types. The hole may be a function that would suggest that a vibration is occurring. Each Calabi-Yau space could have more than one hole, and each hole could exhibit more than one vibrational energy. So we see that the holes in Calabi-Yau shapes—and, further, different shapes have different numbers of holes—directly affects vibration. It may in fact be the hole that is the location of the windings of Planck units and the manifold itself is more of a skeletal structure through which they exist.

We now have a host of manifold shapes with a variety of vibratory positions and a variety of potential vibratory harmonic energies. By plugging various values for holes, specific vibratory patterns result. Some of these vibratory patterns predict, in mathematical terms, the mass of known or theorized subatomic particles. Various numbers of holes allow "families" of vibrational patterns. The three-hole variety seems to match up to the suspected standard model family list of particles and the vibrations that several of these three-hole manifolds support would predict the masses of the standard model. This was very powerful support for the existence of vibrations, strings, and extra dimensions.

As you could imagine, the equations of string theory are exceedingly complex and as it stood, were not easily solved. Many were approximations, albeit approximations to a very small degree. Even so, there was an unsettled nature to these imperfect solutions, which made the theory vulnerable to its critics. Enthusiasm waned as it appeared that string theory was floundering. Articles were even written in the popular press about the impending demise of String Theory.

At the center of the difficulty was the fact that by the mid-1980s there were many interpretations of string theory, some of which seemed to contradict the others. There were five main camps of thought, stemming from the fact that a concept called supersymmetry could be incorporated into string theory in five separate ways, each with its own explanatory framework. This discordance further added to the frustration and lowered the confidence level that they were on the right track.

Edward Witten, the first theoretical physicist to be awarded mathematics' Fields Medal, spoke on this issue in 1995. In his work, he was able to show the interrelationship among all five subtheories and combine them into a single theory that he called M-theory. He explained that the other theories were correct, but that each was limited by specific restraints, and so each described various circumstances but did not provide universality. In this process, it became evident that rather than six extra dimensions (ten total, including time) there were at least seven (eleven total). Not only did this allow for the integration of all of the previous work but also, instead of approximations for results, the new structure allowed for exact results.

All of this was quite satisfying and led to a rebirth and renewed interest in string theory, referred to as the second superstring revolution.

The successful math of M-theory and its power to combine previous theories is very compelling. It is therefore highly likely that it has some role in the overall structure of the universe. It was once hoped that it would answer all questions and be the final theory of everything. It is most probable that it is the missing link between the finest structure we can measure at Planck length and the particles that eventually form matter as we find it in our world. Unfortunately, at this point we get lost in the folds of a seven-dimensional manifold. So far, these hidden convolutions are way beyond the bounds of our understanding and detection.

Addendum: As we are discussing extra dimensions and supposing their whereabouts, it might be mentioned that as part of M-theory, it has been suggested that some of these dimensions are so large that we cannot visualize them from our "antlike" position in our baby universe. To continue, that our universe and its parent universe and grandparent universe…exists on these enormous spatial planes and that our universe's children, grandchildren, great-grandchildren…universes likewise exist on spatial planes unavailable to our inspection.

But still, we may need to reserve some of these dimensions for our smallest spaces.

## *Chapter 7*

# THE MEDIUM BANG THEORY

We are developing an understanding that there is some atomization of the structure of our universe. We are seeing that this medium is more than just the scaffolding through which events happen but the actual sole participant in every one of these events. We have seen that there is a tempo to this that is strict, and follows rules, and makes sense. Chapter 8 takes a great leap of faith. Let's explore the platform upon which this model is structured and the process through which it has evolved.

To begin, the idea that time didn't exist before the Big Bang is just silly. It may be reasonable to say that time is not important in some specific and limited theoretical paradigm, and that a theory will only consider time after a specific event, but to say that there was no moment prior to the Big Bang is nonsense. It feeds our science-fiction appetite but is contrary to the natural order of things. That being the case, if there was time before time zero, it would further imply that the Big Bang was not the beginning of things but rather represents some cardinal event in the history of things. In other words, the Big Bang is probably an incomplete theory. Maybe it is one of many recurring bangs. Maybe it is the biggest of them all or maybe not. Maybe it is one of an infinite cycle of nearly identical events or maybe a unique event that met the specific energy requirements to sustain a flat growth. Maybe it is one of many events, each of which is responsible for the creation of its own universe separate from us.

Some specific event might have occurred fourteen billion years or so ago. This could be that moment that fits our present model of the universe. The Big Bang model can be used to explain the physical behavior and existence of things, but it is nothing more than a model. It is the closest fit for a particular approach to explain the universe mathematically.

As we study our model, we can make some amazing assumptions about the earliest moments. As we approach that first moment we reach intense temperatures, mass, gravity, and pressure. Like calculus, the Big Bang seems to approach but never reach time zero. It approaches but never reaches singularity zero size and infinite mass. It is not necessary for this extreme moment to have existed. It is likely, however, that some moment of some degree of intense temperature and mass happened. The universe probably banged in some sense but criticality did not require true singularity. Only the math requires this singularity. Only the math needs T zero. Of course time happened before T zero.

Taking a look at the Big Bang model, it was theoretically a sudden event. As the pressure, density, and temperature approached infinity, it seemed explosive. Explosion may be a falsely descriptive term, as things didn't get blown apart. They expanded in an even but rapid manner. In a sense, the distinction of whether this was an explosion or a hot sudden expansion may be a moot point, but it does help our mental picture of the behavior of its contents: the contents which are in some manner still expanding to this day.

As mentioned, it would be quite logical that there existed a pre-Big Bang state of the universe because the beginning of time is not so logical. Most of us have wondered, if the Big Bang had been a theoretical first moment, then what was just before that moment, and for that matter what was fifteen billion years before even that? It is conceivable and even likely that something existed before that expansion. It is possible that the preuniverse was a pure uniform "stringy" energy. Enough energy to account for all the energy of today's universe including the mass-energy equivalent of all the stars and all the galaxies times the speed of light times the speed of light. The present known mass represents a small fraction of that original energy. This preuniverse may have had no particular character to it, void and without form.

Einstein gave us one of the ultimate universal truths. Mass and energy are equivalents. The equation $E = mc^2$ does not tell us that mass and energy are somehow related. It does not suggest that somehow mass can be magically transformed into energy. It tells us that mass is, no more or no less, energy. Not that they are interchangeable but are the same thing. Or better put, they are two manifestations of some process or entity. These two manifestations with which we are familiar may only be a subset of possible manifestations for whatever this essence is that makes them up. All of this had to be part of that moment.

It would be a worthwhile field of study to consider is how much or how little of the universe was necessarily bundled in the moment of the process of creation. In the classic Big Bang, the entirety of the universe was contained within that solidarity. In this particular model, everything that is the universe would have been in that moment. I hesitate to say "in that space" because a singularity does not exist in a space that is like any space we can comprehend. For that reason, even though we picture an infinitely small space it is not clear if we could consider any space outside of this singularity primum. This leads us to another big question: if the total universe was entirely contained within the singularity, was there an empty scaffold awaiting the expansion to fill it?

We have a scenario of possible events after the bang, but what was this stuff prior to that critical moment? In the cyclic universe model, an era of expansion reaches its limit and then retracts until the big crunch smashes it all into the singularity, which signals the start of the next expansion. Each limb of the cycle could theoretically take about sixty or seventy billion years and reoccur over and over indefinitely.

The baby universe scenario, with its many subtheories, in general suggests that there have been many universes and that within each generation of universe, a new seed forms. It expands in a way comparable to ours and eventually develops its own pocket or pockets of new babies.

The ekpyrotic model suggests that at some discrete moment in time two separate universe branes collided. That was the Bang moment. The kinetic energy from this collision is the impetus to create the universe. The uneven, nonhomogeneous nature of the event would promote a nonhomogeneous universe better allowing the creation of stars, galaxies, and other imperfections.

What if the fabric of the universe moved in a random fashion as all natural systems do, a wafting universe? Picture possibly a big floppy substrate, twisting and heaving from its own unstable size, the great milieu superstructure. At some moment, that random motion results in a catastrophic twist resulting in an area of knotting like the twist of a clown's balloon animal. This puts the Planck units, which are normally uniformly spaced out, now into a stressful proximity. A significant portion of this fabric knot gets crushed down to criticality.

Over time, pockets of turbulence and instability develop as energy waffles around and interacts. The result is these moments of nonuniformity. Possibly a moment of separation of positive and negative charge occurs or areas of relative concentration or configuration. One disturbed area affects local areas and before long (billions of years), there is a complex storm of Brownian movement. Every right-handed swirl of movement would have a corresponding leftward swirl in order to sustain the process of symmetry. In the great expanse of the preuniverse, the variety of swirls and twists become ever-more complex and numerous, making up every variety of possible configurations. Pure energy is congealed into tiny discrete packets. Each packet steals from the fabric of the universe resulting in nonuniform areas, with their associated gravity as well, allowing them to clump. Clumping would be promoted, as there has been no bang to accelerate them apart. Over the ensuing almost eternity, enough of these pieces and glue gathered together to develop toward critical "mass." This primordial mass would have been very different in character from today's mass, as there was no explosive energy yet to create the quark and boson building blocks.

Soon, in astronomical terms, the amount of mass, energy, pressure, and gravity of this singular mass no longer supports itself. This dark matter would be too big to overcome its self-gravity. These objects collapse into themselves, becoming a black hole, an almost singularity. It is likely that this mass/energy/gravity has its upper limit. Just as a nuclear bomb has its critical mass, beyond which its fission is inevitable, so would the universe. As the crunch preceded, it would most likely reach a critical mass where repulsive temperature/density/pressure outstripped the attractive mass/energy/gravity. At this critical moment, the compression and twist violently come undone with intense pressure and temperature. Bang!

From a theoretical and mathematical standpoint, a pure singularity primum would have been more ideal but was most likely this was not the case. A mechanism for that degree of initial existence would be difficult to fathom. Possibly before our time there was some sort of Big Crunch, as is postulated in a cyclic universe: gravity may have participated in this process to draw everything into proximity, but including every last scrap of everything is unlikely and unnecessary. No matter what model, a critical mass was needed to cause this process, but this critical point would likely be reached before an absolute singularity formation could be reached.

Even though a complete singularity was not reached, the critical mass of this proportion would be similar in character to those moments just after the Big Bang, just as calculated. It would need to reach those high-energy levels for baryogenesis. A significant amount of the universe would necessarily be in this crunch, a critical amount, leaving untold space outside of this event.

We now have arrived at the beginning of our story, the great expansion that continues to this day. We see that the galaxies are moving apart as per the red shift of Hubble, moving at nearly the speed of light toward the periphery of the universe. This is an optical illusion. The universe is expanding like a sheet of rubber. The galaxies are embedded into the sheet of rubber and remain stationary in respect to their position in the rubber but appear to expand as they are part of the stretching process. They are like the funny paper image lifted onto silly putty. They are part of the stretch of the universe, not objects moving through the scaffolding.

In the decades to follow general relativity, an exact solution was determined for Einstein's field equation assuming this homogeneous atomized universe and is called the FLRW metric. Named after Alexander Friedmann, George Lemaitre, Howard P. Robertson, and Arthur Geoffrey Walker, it is a useful view of the universe. Although the solution may seem exact in its value, it is an approximation of the universe in that it doesn't really account for the tiny bits of stuff out there, like galaxies. On a cosmic scale these are of minimal concern. In any event, we see in these equations that the quiet empty space out there has some sort of underlying structure.

Cold dark matter (Lambda-CDM model) may account for approximately 27 percent of our universe's energy-matter. This model does account for the lumpiness of galaxies as well as the empty background space with its inherent energy as demonstrated by microwave radiation. It validates the cosmologic constant and the expansion of the universe (actually accelerated expansion). In reverse, it suggests the Big Bang origin of the universe.

One of the most important topics in current physics research, and one of the most mysterious of physical concepts, is known as dark energy. In order to explain the theories of an expanding and accelerating universe, the total calculated density of the universe and the behavior of large-scale wave patterns of mass of the universe, some other untouchable property must exist. It is different from common matter and energy. It is different from dark matter. It has been dubbed dark energy, but its true essence is, thus far, unknown.

There is very little dark energy in any given space with a density of $1.67 \times 10^{-27}$ kg/m$^3$. At this density, dark energy is unlikely to be detectable in laboratory experiments. In the entire expanse of our solar system, there is only about six tons total. In all of space, however, there is a whole bunch of dark energy. It fills the entire expanse of space in a uniform manner. Despite its very low density, dark energy actually makes up most of the density of the universe. It's a big universe!

It is speculated (calculated) that ordinary matter of stars and planets makes up 4.9 percent of the total mass-energy equivalence.[50] The dark matter of black holes and other such scary stuff accounts for 26.8 percent. The remaining 68.3 percent of all that is, is dark energy, at least in theory.

When considering dark energy, we cannot help but notice conceptual similarities to our concept of Planck units. Here it is once again helpful to look to a concept first put forth by Planck himself, which now frequently pops up in discussions of dark energy: zero-point energy. Zero-point energy, at first an apparent mathematical oddity that came from Planck's work on his radiation law, holds that an object will still have some energy, even at a temperature of absolute zero. In discussing Planck's legacy, Brandon Brown notes the potential application of zero-point energy to theories of dark energy: "The zero-point energy of the cosmos is automatically a candidate for determining the source of 'dark energy'...The dark energy is still mysterious as of this [2015] writing, and it is not properly accounted for by any computation of zero-point energy, but the two often appear in the same conversations." Could it be that the zero-point energy is the most basic level of energy: that of a single Planck unit in the deepest reaches of space? And could it be that the aggregate zero-point energy from this collection of cold, lonely, quiescent Planck units in the deepest reaches of space is what we know of as the mysterious dark energy?

In his book *From Eternity to Here*, nearly a century after Planck put forth the idea of zero-point energy, the physicist Sean Carroll describes the properties of dark energy in terms startlingly similar to our proposed properties of Planck units. Carroll notes that the two things we do know for certain about dark energy are that it is constant throughout space and nearly constant in density through time. Like Planck's zero-point energy, there is "a minimum amount of energy inherent in the fabric of space-time itself"[51] that is undetectable and "unusable" because it exists in such a high-entropy (that is, completely and evenly distributed) state.

Dark energy is the most abundant thing, and yet we have no clear notion of it! Because we have defined energy by our common experience, it seems that energy should have a source and behavior. It shouldn't just sit out there quietly. Here again, in dark energy, we have a theory that suggests an all-pervasive entity that fills all of space, that there is a "seething and invisible lake of energy"[52] that pervades space itself. This is Einstein's cosmologic constant as an entity rather than a mathematical construct. Nearly every contemporary accepted physical theory acknowledges the existence of dark energy, but none of these theories makes the leap to recognize that it is also the long disavowed luminiferous aether (though with an expanded, twenty-first century definition).

We now have our scaffolding and the chunks of matter, and Albert Einstein tells us how they behave. The theory of general relativity is a generic equation that is applied to sixteen different coordinate combinations. The sixteen equations are based on a four-by-four matrix relating the three traditional spatial coordinates ($x$, $y$, and $z$ axes) and time. In this four-by-four matrix, we can exclude the four noncomparison equations ($x$ by $x$, $y$ by $y$, $z$ by $z$, $t$ by $t$), leaving ten equations.

The theory of general relativity is this set of equations. It can tell us various things, depending on our specific interests. For our purpose, it would be appropriate to say that it is a comparison of how mass warps space-time coordinates or how space-time defines how mass moves (bends). The sum of effects of large mass behavior is very complex, but this set of ten equations allows an attempt at breaking down the mechanisms in simpler terms.

To take a step back, we study geometry on a flat sheet of paper, Euclid gave us the rules, and they behave beautifully for most everything we need in daily life. We can ignore space-time curvature and gravity and get by quite well. However, everything we do is in curved space-time and under the influence of gravity. Our measurements depend on the "flat" Earth's surface (our traditional $x$ axis) and the "upness" and "downness" we feel on Earth (the $y$ axis). To account for the curvature and our arbitrary position in the universe, we would use Einstein's field equations to adjust these simple vectors for these influences. These apply pressure (tension) in an organized and predictable fashion. By means of these tensors we can account for the differences between expected Euclidean behavior and geometry on an invisible but curved surface and also the degree to which these pressures affect everything.

In 1916, Einstein published his theory of general relativity. He did so in the setting of the general consensus that the universe was static. Gravity, however, would not allow a static universe, as the gravitational force would tend to cause a contraction and collapse of the universe. The mass of the universe should pull everything together. To keep the universe from collapsing, at least mathematically, he needed to add an interesting fudge factor that he named the cosmologic constant. It applies an opposing outward pressure equal to the inward tendency of the gravity of the universe.

In 1928, Hubble showed evidence that galaxies were retreating from us, thus suggesting an expanding universe. This paradigm shift in thinking was amazingly well accepted and adopted but made the fudge factor no longer necessary. (Side note: In every book on the subject, it is obligatory but cliché to say at this point: when Einstein understood that he did not need his cosmologic constant, he described this as the biggest blunder of his life. I'm glad I got that out.)

Nonetheless, the cosmologic constant does describe the opposing energy/pressure that would be needed to exactly balance the mass-energy/gravity of the universe in if it weren't accelerating. Even though it is not for the purpose of keeping the universe in check, the cosmologic constant does happen to describe that matrix that makes up the universe. It now stands on its own as the cornerstone of one of the theories of what is out there in the darkness accounting for the not-quite-nothingness that is space.

The cosmologic constant defines an energy amount. It is the summation of all the individual energy amounts of all the quanta of the universe. This energy can also be calculated at its individual quantum level. Energy E equals the Planck constant $h$ times the frequency (of a photon for instance) μ. $E = h\mu$.[53]

More recently, loop quantum gravity explores the idea of these individual quanta in space. With so many involved in LQC, there are necessarily many interpretations of the meaning. In general, most would admit to an atomization of the universe and that it exists at Planck scales. Most would say that on one hand they live under the rules dictated by Einstein, which state that there is no fixed background upon which to localize things and events, that position and time are relative. On the other hand, there is a smallest space out there in which all things exist and happen.

LQG is the closest description thus far to the mechanism by which the universe works. Unfortunately, they stop short of these spaces existing out there as aether, and they still expect light particles to be whizzing around.

This issue can be resolved in defining background dependence. While the events are enmeshed and dependent on the underlying atomic structure of the universe, there is no point that a photon or particle or process owns that space. They pass through the Planck units instantaneously and seamlessly with no lasting effects. Planck units remain unchanged and would not have any stake in participation of any event. All Planck units are essentially identical and featureless, and so to identify a specific space would be meaningless. In this way, aether can exist but not impart any meaningful structure and reference. This would allow for both aether and for Einstein's equations to continue to function properly.

One winter day I caught some virus, and by 4:00 p.m., I went to bed. I was very ill that night—in long johns under sweat pants under four blankets, I was still chilling. At one point I thought I might end up in the hospital, but I finally went to sleep. I woke up around 12:30 a.m., and my fever had broken; and just like that, I felt almost back to normal. Strangest illness I've ever experienced. But now, it's 12:30 a.m., and I'm wide awake, already having slept a whole night's worth of sleep and with all of the previous day's events washed from my brain by a febrile illness, and my mind wandered to the most recent book I had been reading, Richard Feynman's *QED: The Strange Theory of Light and Matter*.

In the paperback copy that I had, on page 129, was a paragraph that I had seen quoted elsewhere, and more than once. It goes:

> There is a most profound and beautiful question associated with the observed coupling constant, e—the amplitude for a real electron to emit or absorb a real photon. It is a simple number that has been experimentally determined to be close to –0.08542455. (My physicist friends won't recognize this number, because they like to remember it as the inverse of its square: about 137.03597, with an uncertainty of about two in the last decimal place. It has been a mystery ever since

> it was discovered more than fifty years ago, and all good theoretical physicists put this number up on their wall and worry about it.)

I don't have this number on my wall, and I have never worried about it, but this warning seemed strong enough that I thought I should look into this a bit. This number, 1/137.03597, is called the fine structure constant, FSC, or Sommerfeld constant after Arnold Sommerfeld who, in 1916, described this recurring, dimensionless value denoted by the lower case Greek letter alpha ($\alpha$).

It immediately reminds us of another interesting number, pi. Pi is the ratio of the circumference of a circle to its diameter. There is no measurement of circles, spheres, curves, or angular motion that do not depend on this number. As a ratio, it is unitless (i.e., dimensionless). It is an irrational number in that it cannot be represented as a ratio between two integers and therefore it never ends in a final digit or repeating sequence. Moreover, it is a transcendental number in that it is not a root (or solution) of any algebraic equation (i.e., a nonzero polynomial equation with rational coefficients).

FSC shares some of this mystery. It is a unitless number that can be calculated in a variety of ways as a ratio of various electromagnetic events, and yet it is not clear what this number may represent physically. The FSC is a value defined by the following:

1. Measures the strength of the EMF, which controls charged elementary particles

2. The probability (amplitude) that an electron will emit a photon
3. The square of the elementary charge over the Planck charge
4. Ratio of two energies (the energy to overcome the electrostatic repulsion between two electrons separated by a distance of d over the energy of a single photon of wavelength $\lambda = 2\pi d$)
5. Ratio of velocities of electron in first Bohr orbit to speed of light in vacuum
6. Coupling constant: the strength of interaction between electrons and photons

As mundane as these descriptors may be, it turns out that this number is responsible for the stability of all that is. If the FSC were even the tiniest bit higher or lower, the atomic structure of the elements would be enormously bigger or smaller. Anything more than a tiny change might not allow atomic structure at all. So, it turns out, our existence depends on this number.

A brief digression to a story told often in physics books about the cosmic microwave background radiation (CMBR). Bell Telephone Company had a state-of-the-art radio communication antenna on Crawford Hill in Holmdel, New Jersey. It had a large surface horn detector dish to connect with the echo satellite along with very precise electronic monitoring equipment. Two radio astronomers, Arno Penzias and Robert Wilson, used this antenna to measure the intensity of radio waves bounced off Echo balloon satellites. To do this, they had the difficult task of eliminating background noise so that they could specifically study our emitted waves. This turned out to be a much bigger chore than they expected. They used some specialized techniques to isolate electrical noise inherent in the system. They famously went out to the antenna to scrape the copious layers of pigeon droppings. None of these efforts alleviated the constant hum at 7.35 cm wavelength (which falls in the microwave portion of the spectrum). Through several other assumptions, they concluded that this was not coming from some focal emitting source but was more consistent with random waves from a general, universal origin. Based on work with blackbody radiation, it is known that the frequency of an emitted signal is proportional to its temperature and that at this frequency (given a few more assumptions) the temperature would have been in the neighborhood of 2.5 to 4.5 degrees above absolute zero. Not that they assumed that this radiation equated to a temperature, but only that this would approximately be the temperature equivalent of their findings.

Dicke, Peebles, Polli, and Wilkinson were approaching a similar topic from another vantage. In order for the presumed Big Bang to be successful at enormously high temperatures, there needed to be a way to account for why all of the hydrogen and helium did not get cooked into heavier elements. This presumably was due to enormous amounts of radiation present at the time. Calculating the amount of radiation needed at the theorized temperatures of the first few moments and the present expansion of the universe with its dilution of that radiation, the fossil, leftover radiation should be approximately in the range of ten degrees above absolute zero.[54]

We think of microwaves as moving, and yet where are these microwaves coming from or moving toward? In the Big Bang scheme, there should have been a point of origin and possibly a wave front at 13.7 billion light-years from that region, not a diffuse soup of waves traveling in every which direction. These are not stationary waves of static electricity; instead, they may be the static framework of Planck units. This universal experiment of blackbody radiation raises the question of what this radiation is doing out there. In calculations it may correspond to microwave, but it is only human nature to label it as something we have studied. This is not random microwaves, and is not leftover temperature, but the stationary quantum jitters of Planck units.

I have told the story of absolute zero temperature and absolute zero distance. These numbers are the absolute minimum values allowable by our universe given the dictates of the atomic Planck structure. Certainly we could write down numbers smaller than these values, but they have no particular essence in our present universe at its present degree of expansion.

Out in the bitter cold of outer space at –270° C, the poor, lonely Planck units are quivering. This is the absolute zero energy state of a quiescent Planck unit. This zero energy is certainly too small to measure for any given unit, but collectively, they make their presence and their inherent resting vibratory energy evident. Temperature, microwaves, and dark energy are all manifestations of this quivering.

Quantum theory tells us by its very definition that all things occur in discrete steps. The next quantum energy step for a Planck unit is about 137 times greater than its absolute zero energy resting state. If at rest it were actually at some true zero, with no intrinsic energy or vibration or whatever its energy is manifest, then the ratio of lowest energy to next quantum minimum amount would be 0/137, or zero. The ratio, however, is a nonzero number, 137, suggesting a basal nonzero resting energy. At the single Planck level, this would be difficult with our meager instruments to measure, but we see this ratio demonstrate itself over and over as we look at energy behavior of bigger chunks of Planck units such as those seen in an atom of hydrogen.

# *Chapter 8*

## MATTER

This world we think we know so well—we have examined and categorized it thoroughly. In the past four hundred years, we have moved from four basic elements (earth, wind, fire, water) to molecules and their constituent atoms to a zoo of subatomic particles. Even in this Holy Grail, standard-model list of subatomic particles, there are still mountains of unanswered questions. Particles with no mass? Duality of photons? Gravitons? Dark matter? Higgs? There are lots of weird things that we need to take on faith to make the standard model fully functional.

So what's the matter? Everything is matter! (Bill Nye.) We assume that matter just *is*, just as we assumed a few hundred years ago that things fall to the ground because "down" just *is*. When Einstein defined matter as having an energy equivalent, it suggested matter might have another essence. It raises the possibility that matter doesn't exist exactly how we think it does. That maybe it is no more than smoke and mirrors designed by nature to fool us into believing in its existence, our existence. Possibly it could be a special effect that occurs when some generic building block has an enormous dose of energy applied until it appears recognizable to us as in the form with which we are so familiar.

In nature we see a recurring pattern of reductionism. Very complex systems are made up of simpler and simpler building blocks. Plants and animals are made of cells. Cells are made of organelles; organelles are made of fats, proteins, carbohydrates, and nucleic acids. These macromolecules are made up of a dozen or so elemental atoms. Atoms, once thought unsplittable, are made up of much simpler protons neutrons and electrons. These subatomic particles are theorized to be made up from the list of ingredients of the standard model. It would not be nature's way to end the subdivision process with a list of ingredients. The "fact" that there is a list would suggest that there must be at least another level of simplicity, if not more, until we arrive at one essential simple common thing.

Wouldn't it fit that just as a pine tree or a human can be broken down into simpler and common units that so, too, quarks and gluons can be further broken down into their more basic units? Wouldn't it also fit that at the moment of creation of these particles, they originated from a more generic substrate?

Superstring theory (string theory for short) may be a bridge to this end. String theory has powerful mathematical tools to help explain the universe. At the core is the belief that stringlike, one-dimensional units are the basic building blocks of matter. Strings are theoretical at this point. They have not been assigned a size, mass, or energy. It is suggested that a single string may make up a quark or other elementary particle, and the particle's uniqueness comes from the intrinsic harmonic vibration of a specific string. In this respect, string theory may represent the next layer beneath the standard model. It may assist as the explanatory mechanism behind the standard model's descriptive framework, and yet string theory remains incomplete.

In standard string theory, many consider the string to be the most simple, generic unit. It is felt that the string itself has much less mass than a quark, or is massless. Strings or something like them may be a reasonable intermediate step to forming mass, and string theory explains the mass of an elementary particle as a reflection of the energy of the vibrational pattern of its fundamental string, thus directly incorporating Special Relativity's connection between energy and mass. More broadly, going beyond just the property of mass, the theory holds that "the observed properties of each elementary particle arise because its internal string undergoes a particular resonant vibrational pattern."[55] However, string theory is far from a final, be-all and end-all physical theory. Brian Greene himself notes that it may just be that strings "are one more layer in the cosmic onion, a layer that becomes visible at Planck length, although not the final layer. In this case, strings could be made up of yet smaller structures."[56] Let's run with that idea and explore the possibility that string behavior is actually a calculable manifestation of a more basic unit, our now-familiar Planck units.

In every natural system, no matter how complex, the essential building blocks tend to be very simple. Nature takes the simplest units and by random chance and the expanse of time sends them in a nearly infinite array of configurations. These configurations are almost never favorable or stable and so disintegrate into oblivion. But every so often, a random event leads to a more stable configuration that has staying power.

For instance, consider the primordial soup that evolves eventually into us. In this natural system, the essential building blocks floating in the vast ancient oceans were simple atoms. Volcanoes erupt, lightning strikes, gases spew; and in this process and entirely by chance, nature experiments with various molecular combinations. In these oceans of experimental chance, molecules are of little benefit in the progression to biological material, but again by chance and the great expanse of our prehistoric ocean and over the course of billions of years, stable, very simple nucleic acids happened. These interact in nearly infinite ways with other molecules until, by chance, an even more "valuable" molecule might form. This process, although very inefficient, continues over hundreds of millions of years until, accidentally, a small, replicating single strand of molecules forms. And so ends the hard part of our history.

Nature repeats this pattern in every system. Complex stuff does not form de novo. Complexity comes from a random, prolonged process of ordering.

Picture a nice, fine-gold necklace. Place it in your pocket for a couple of hours, and when you reach for it you will find a rat's nest of knots and twists. What happened to your necklace by chance is now going to take you all afternoon and a lot of your energy to undo. A significant amount of energy was actually put into creating this mess as you walked around all day. Many miniknots formed and unformed in the process. Some knots were more stable and persisted. Some were tightened into even more stable configurations that will take a lot of work to undo.

The problem with this analogy is that the knots occur in a random fashion. They occur in every possible pattern. Our earthly particles seem well organized and come in limited configuration options. Of the many atomic knots that formed, only a very few had enough value to be sustainable and usable.

For the sake of this discussion, let's assume the making of these knots comes from the twist of "strings." Strings in the front pocket of the universe get formed, pushed, and twisted in infinite directions with an infinite variety of energy amounts. Some energy is barely able to jiggle these minuscule strings while maybe a little more is enough to make a string fold back on itself. This simple fold is too unstable and so snaps right back into its more stable natural baseline configuration. Then some plasma level energy event occurs, such as a supernova, black hole, or Big Bang. This energy input bends and twists the heck out of these strings. Most of these will not twist into anything stable or sustainable, but just by chance, some of them find that perfect knot that does not readily come undone. Undoing this knot would take an enormous amount of energy, so, in a way, the energy of the plasma event is stored in this knot. In the string world, these knots have only a few stable configurations. These configurations correspond to those higher levels of dimensionality in the superstring mathematical scheme.

If you were to imagine a string, you would necessarily picture something with a length greater than its width. Strings, therefore, could not by definition be the smallest possible thing. The smallest thing has Planck length. If a string were Planck length, and Planck length were the shortest possible distance, and its short axis was also limited to Planck length, then we are defining a point. The Planck unit is the smallest space measuring Planck length in every direction. Strings could, however, be collections of these simpler units. Any number of individual units could make up a string. If we were to expand our definition of strings to include every possible manifestation of Planck behavior, then we could consider the smallest simplest string as be a single, quiescent nonvibrating Planck unit (even though that single dot would not be very stringlike). A vibration of these units translated down the line yields the next simplest string, the electromagnetic wave (or photon, if you still like to pretend it's a particle).

In essence, a string is not a specific thing but instead the behavior of a collection of units. A string exists only as long as that function exists. A light wave is stringy behavior until that energy is absorbed somewhere down the line. A particle is a string or collection of strings twisted into very specific patterns.

Strings and/or branes may be the manifestation of a group of units disturbed from their most diffuse, lowest energy state. These groupings become evident in their more disturbed vibrating state or folded high-energy, massive state. **Everything is Planck units.** Planck units are more than just the background structure through which matter moves, they are the actual matter itself. All particles, from "photons" to protons to galaxies, are complexes of Planck units.

Strings can take up as many Planck units as necessary—but at a cost. The more complex they become, either the more unstable they are or the more energy is required to get them to the next stable configuration. Just as with our primordial soup, it may be that in the earliest moments of our universe, a nearly infinite number of random string disturbances occurred. Rarely (only enough to create all of the mass of the universe), these twisted strings settle into a stable configuration. It is possible that these more stable configurations are either only possible or at least only usable in the patterns defined by the math of string theory.

The singularity of the Big Bang presumably had mass. As a singular event, the Big Bang set in motion the creation of present-day particles, but that doesn't mean that the primordial mass was necessarily made of the same components. This might suggest a pre-Big Bang era of random mass building, not our present-day neutrons, protons, and electrons but primal strings of premass. As random mass built up in a non-expanding closed universe, it would tend to collapse on itself until reaching a critical mass/density/temperature. Whatever parts of the standard model that actually exist might be this universe's iteration of particle options.

With the intense temperature available at $T_0$ + two minutes (the first couple minutes after the Bang), all sorts of configurations would occur. Most of these configurations would not be stable. They would be slip-knots, and all that available kinetic energy would destroy them as quickly as they formed. But some would find a more stable configuration. Just like a necklace in your pocket, it would take a great deal of energy to untangle that knot once formed. These knots tend to settle into the various levels of complexity, possibly defined by Calabi-Yau math. These various levels of complexity are very reminiscent of the extra dimensions found within the mathematics of string theory.

This creation story is in vast variance with the traditional thought that matter existed at $T_0$. This story does not require the entire universe to exist in a singularity. Instead it does require an enormous amount of some form of matter, at critical density. $T_0$ is not necessary and a true singularity is not necessary. Good thing, as neither a gathering of every speck of everything into one ball or a true moment of $T_0$ are at all logical.

Even so, as we covered in the previous chapter, there was most likely some cardinal critical event that was Big Bang-ish. And with the fire of that first furnace we have the stable formation of strings (or swaths) of Planck units ready for further manipulation into our present-day recognizable particles.

Imagine a two-dimensional sheet of Planck field. Add energy to this system. Add so much kinetic energy that violent twists occur. These kinks in the sheet take every conceivable shape and usually just fall back to their simple starting point. But some twists result in a knot that is much more difficult to undo. These knots take a great deal of energy to create but, once formed, are nearly impossible to unknot. This energy can now safely exist in these knots as potential energy and can be quantified by $E = mc^2$. I use a two-dimensional sheet in this model for simplicity, but we must remember that these twists actually involve a volume of Planck units.

Strings are a generic collection of the most basic unit. Strings vary in their manifestation but are made entirely of these most basic units. The most basic unit is defined by a Planck units constant. The Planck constant is firm and fixed for a reason. It defines the smallest possible dimension. As $c$ defines the universal speed limit, $h$ defines the universal length limit.

In order for a string to be stringlike, we would picture a series of units to give it length. As mentioned, it would not be unreasonable to consider the single unit at rest to be the absolutely simplest string. This is that single unit surrounded by all of its neighboring units in the deepest, coldest outer space. Not interacting with any other unit. Not participating in a vibratory event. A string in its quiescent state is the essential unit of dark energy.

The next level of complexity occurs when a unit transmits energy in the form of a vibration from its neighbor to another neighbor, as is seen in the transmission of an electromagnetic impulse. In this schema, the string could be considered the entire path of units transmitting the electromagnetic wave. The actual string could not be defined, because, as mentioned in chapter 5, an electromagnetic wave front goes off in all directions. The string in this case would be the units that by chance happened to create the path from source to detector. Again, this is why it might be more appropriate to call this event string behavior rather than an individual string. The front goes in an infinite number of paths. Each one is stringy and vibratory, but just as trees falling in the woods, only the ones we detect seem to be real and interesting to us.

The empty space of the universe has a static energy defined as the cosmologic constant. This energy is the same energy of the Big Bang. Spread out over such a vast expanse, this energy seems quite innocuous, but the total energy is immense. It can still be visualized as background microwaves and measured as a couple degrees above absolute zero. Collectively, this is what we consider to be dark energy. It also happens to be the Higgs field. In the hyperdense first moments, this bland, featureless field was manipulated under the intense heat and pressure into a hyperenergetic form: plasma. As this cooled, as it passed through various crucial energy levels, some of the plasma settled back into its original low complexity, empty space state, whereas some of it was twisted, folded, and spun into ultrastable but much higher potential energy states.

These stable knots would favor specific patterns and are expressed mathematically as the Calabi-Yau manifolds of string theory. We now have three distinct levels of string behavior: the individual quiescent units that collectively make up outer space, a path of vibratory units making the temporary string of an electromagnetic event, and a knot of units making a manifold of matter.

On the far end of the scale, the strings making up matter could be of enormous size. This is what required some sort of Big Bang event. As a swath of individual units was subjected to enormous levels of kinetic energy, they got twisted and pounded into every possible pattern. With complexity comes enormous energy expenditure. Most twists would likely not result in a stable configuration. Other possible configurations would have billions or trillions of these units folded into very specific, stable patterns, forming a single boson. It is almost certain that these stable configurations correspond to the extra dimensions predicted by string theory. The baryons of the standard model could represent the larger side of the scale of stable string formation (assuming that they actually exist as predicted).

A stable string formation would have to meet some prerequisites. Not just any old knot will do. This next level of complexity must retain its pattern despite its natural tendency toward entropy and the release of enormous energy. Quarks could represent this knot.

All quarks would then be strings, but not all strings, obviously, are quarks. Strings are not permanent structures but instead represent the various behaviors of these individual units as they participate in everything that is. Strings come and go at the speed of light. Strings are everything and yet never actually have a true existence of their own.

So here we are with a jar full of quarks giving off their electromagnetic field, their gravitational field, and interacting with the Higgs ocean to impart mass.

First of all, the Higgs boson needs to go away. It is postulated because, in theory, every particle has an associated field and every field describes a particle. For the Higgs field to exist, then, there ought to be an associated particle. It was the particle that made Higgs's equations especially different from the other teams working essentially on the same process. Though it is true that every particle and every event does have a field, they all share the same field, the one and only field, the ocean of Planck units.

This Planck ocean is the entirety of the universe. It is the connectivity of the Planck units. As Planck units are swept up into a Calabi-Yau configuration, the surrounding units are its field. The disturbance of new units entering and leaving this configuration leaves their trace as field behavior. The resistance, as these units trace their path through the complex structure, imparts what we sense as mass.

As mentioned, a string supposedly makes up the entirety of a quark. The mass of the quark, however, is supposedly dependent on the energy of the harmonic vibration of the string, not the intrinsic mass of the string (if it actually has its own mass). Presumably, under the old order, if a quark is to maintain its mass, its string must maintain its vibration for eons. The quark is theoretically much smaller than the particle it makes up. For instance, the three quarks that make up a proton represent about 1 percent of the total mass of the proton. It is theorized that the remaining 99 percent of the mass is a manifestation of the binding energy of those quarks at a rate of $E = mc^2$.

I would suggest that the above theory is not correct. A quark is actually a knot of Planck units. It does not have some unique quark electromagnetic field. It does not have some unique quark gravitational field. The field that a particle experiences is what we have recently been calling the Higgs field. The associated electromagnetic field, gravitational field, particle field, and particle itself are all various local expressions of the great field—the field of Planck units—that is the expanse of the universe. All fields are actually focal alterations of this great field because all fields are various expressions of these basic Planck units.

Conversely, fields do not define a particle. As all particle fields are the same as "Higgs field," they are quite generic. The measurable differences in these fields are related to how each particle or electromagnetic event distort the Planck units. The specific distortion is how a specific particle is manifest. For instance, an up quark depends on its very specific configuration, the winding into its very specific Calabi-Yau space. This winding will have a very specific and predictable effect on the surrounding field. Field theory is the mathematical or diagrammatic evaluation of how things affect the local Planck unit field.

**A particle forms when Planck units are gathered into a stable configuration. It gains mass as that configuration moves in space through a resistant ocean of units.**

If you had a collection of marbles, all apparently the same size, and you could measure them at their finest level, it would be clear that they all varied significantly. It would be unlikely that out of a thousand similar marbles any two of them shared the same number of molecules or mass or shape or impurities. As it is for marbles, so it is for sand or BBs or any other collection that appears to have some uniformity.

Particles are a significant exception. Protons have a mass of $1.67262178 \times 10^{-27}$ kg. There is no exception; there is no range of proton-ness. As precisely as we can measure them, they will always be an exact number. Any variation is related to our precision in measuring them and not in a variation among protons. A proton from the furthest galaxy would have a mass of exactly $1.67262178 \times 10^{-27}$ kg.

We are certainly happy that this is so as it makes theories and calculations simpler and more predictable. It is no coincidence, however. It is due to the limited number of stable configurations that could form in these Calabi-Yau patterns at the available energies of the Big Bang.

Although not important to this discussion, it should be remembered that the total number of particles formed was many times greater than what existed even a short time after their formation, as they were formed in the presence an almost equal number of antimatter knots, which annihilated most of them. The immense nature of our universe is just unfathomable.

A particle accelerator sends one proton in one direction at nearly the speed of light and another in the opposite direction at nearly the speed of light. If one of the protons were fitted with a speed gun, it should see the other proton approaching at nearly twice the speed of light. Au contraire, as these protons travel they are the knot of units. The Planck units are incorporated into the proton only for the briefest time. At the instant of detection, the Planck unit that is temporarily the detector has not been in motion. In our time the detector proton is moving, but at the extreme of Planck time, it was sitting there waiting for that proton. It will share that vibration from the oncoming proton in that Planck space for Planck time. It will measure the incoming impulse as the absolute speed maximum, the speed of light. This is the theory of relativity at its extreme. Either proton can claim ownership of that space for that briefest moment while watching the other proton approach at the speed of light.

The gathering up process of Planck units into a Calabi-Yau knot is the process of matter formation. It does so at the cost of $E = mc^2$. The resistance to this is the process that imparts mass. This is the same resistance attributed to the Higgs process.

The Higgs process is considered to be the resistance that a particle experiences to its motion through space. The resistance is thought to be the process that imparts mass. Every single quark that makes up this particle would experience this field resistance separately, adding up to the total mass of the final chunk of matter. In mass formation, this resistance behaves in a very peculiar way. The motion of a submarine through the ocean requires enormous energy to propel it against the resistance of the water. The Higgs resistance, however, does not slow down the progress of the particle or require power to turn the propellers to continue its forward progress.

This amazing potential occurs because, continuing with the submarine analogy, not only does the water represent the Higgs resistance, but as the water molecules reach the submarine, they get incorporated into what actually makes up the entirety of the submarine. As the submarine moves forward, the water molecules change to steel and then rubber, and then electronics and then sailors until the submarine leaves that space, at which point the water looks like water again, just as it started out. Rather than being excluded from the submarine, the molecules pass seamlessly through the submarine, actually becoming the submarine briefly.

In this analogy, the water molecule is akin to the Planck unit. A universe of Planck units minding their own business experiences the knot of an approaching quark. As the knot passes through that space, the Planck units become that knot. The units glide seamlessly though the twists and turns of that knot, giving it structure. The alterations in the local units cause the tension that gives us the impression of mass.

These Planck units are all there is. They are the Higgs ocean of molasses that everything must move through to gain mass, but at the same time, they are the quintessence of what everything is made of. It is the essential fabric of everything. Every vacuum, light wave, atom, star, galaxy, or person is no more and no less a manifestation of this great field, which is composed of almost infinite and almost dimensionless units.

Sitting motionless, these units account for all the dark energy of the universe. With only a small pluck of energy, the vibration of light is transmitted from one edge of the universe to the other. Folded into unseen dimensions, they impart mass.

Thanks to the complex and beautiful mathematics of string theory, the domain of these extra dimensions has been defined. Every particle of our universe is a swath of this fabric folded, quite by accident, into various stable configurations. An infinite number of configurations occurred but only stable ones persisted. These patterns may coincide with the Calabi-Yau shapes and are predictable, recurring, and abundant. When the dust finally settled, there were several species of possible shapes with various energies and complexity. This is the standard model, or at least the portion of the standard model that is not make-believe. In addition to these subparticles, it would be likely that there are other complex stable configurations out there that do not exist in enough quantity to significantly add to this schema.

As one pictures the local Planck units morphed into various blobs until a familiar pattern occurs, say, a bottom quark for instance, it is truly mind-blowing to consider the amount of energy required in the process. In a perfectly efficient system where all of the needed energy to create this quark is present, that energy would be defined by $E = mc^2$.

An electron volt is the energy it takes to move an electron though a 1volt potential field. It can be used to describe an amount of energy or an equivalent amount of mass. It is an exceptionally small unit.

A volt equals one joule per coulomb. An electron charge equals $1.602 \times 10^{-19}$ coulombs (C). Therefore, an electron volt equals $1.602 \times 10^{-19}$ J.

A mass of a bottom quark is estimated at 4.2 billion electron volts (4.2 GeV).

The cosmologic constant tells us that "empty space" contains energy. Attempts at defining this value are fraught with assumptions. One calculation that at least offers a reasonable value puts the mass density at $10^{-27}$ kg/m$^3$. If correct, this means that a cubic meter of empty space carries $10^{-9}$ J or 62.4150647996 GeV. So to create a bottom quark it would take a twisting and gathering of 67,000 cubic centimeters of empty space energetically hidden in the extra dimensions of Kaluza-Klein world.

If it were only that simple. Whereas this big swath of fabric is tied up into this tiny package so profoundly that we can't get at it, the individual Planck units of this fabric and the fabric itself glides into and out of this package in an instant.

The package defines the particle, but the act of the contortions each bit of fabric goes through to enter and exist and exit the package is what imparts mass. We think of it as a sort of resistance, but this resistance does not steal energy from the system. The process is the equivalent of what the standard model considers a gluon. The $E = mc^2$ of a quark remains the same throughout its lifetime. Resistance should take some energy input to overcome or should lead to a gradual expenditure of the intrinsic energy, but this is not the case. This would suggest that the mass imparting process is not truly resistant.

The particle package—the fabric of units—remains stable. As a particle moves through the Higgs field, the empty space gets brought into the ultrastable folded area, passing through every bit of that complex fold until coming out the other end unchanged and in proximity to its original neighbor. At least on the time scale of particle motion, the fabric stays put while the particle processes though it. The individual units remain stationary in respect to their neighbors. In a sense, the units could be thought of only as they exist as a fabric and not as a solitary quantum.

Though it may be reasonable to consider this evolving string of billions of Planck units flowing into then back out of a single knot of one quark, it's much more difficult to imagine them flowing into and out of successive quarks enough to make up the entire Earth and doing so at cosmic speeds. But then again, the universe is filled with inexplicable extremes.

Mass is a measure of the energy to create a blob and continually recreate it as it moves through the eternal field of Planck's ocean. Mass is relative: when we push through the resistance of this field, other objects moving at a relatively similar speed will manifest their mass. If the observer and object speed up together, the apparent relative mass stays the same. If the object speeds up relative to the observer, its relative mass would appear to increase. As the object's velocity approaches light speed the mass as measured by the observer will approach infinity. Mass is not an actual thing but an observer's interpretation of the event of matter passing through space.

Furthermore, the matter-forming process is a little fuzzier than that. There is a stringy property called uncertainty. Let's, for a moment, step back to electromagnetism. In the propagation of an electromagnetic wave front, the brief change (vibration) that occurs at the individual Planck is transmitted to the next unit. When we picture this transmission it is in the form of an arrow moving from source to target. In actuality, each Planck vibration is transmitted in some degree to each of its neighboring Planck units. The net energy transmission is indeed in the direction of the arrow, but a little of that energy moves to adjacent Planck units. Each Planck unit vibration is essentially the initiation of a new wave front.

For instance, say that of all the energy in one Planck vibration, 98 percent gets transmitted to the next unit directly down the path. That other 2 percent is dispersed among all the other Planck units in direct contact. A big portion of the dispersed energy goes to the Planck unit just off that direct line, but some goes to the Planck unit at 90 degrees, and a tiny bit even goes backward to the unit at 180 degrees. Where this vibratory energy is at any given moment is what defines that specific electromagnetic ray. The dispersion of that energy is why we say that ray is mostly here but a little bit there at the same time. That fuzziness is the essence of the uncertainty principle. To make things even more complex, at the same time, that same unit is receiving vibratory information from its neighbors and with various vectors of influence.

That image is much more evident in the electromagnetic behavior of strings, and we can even picture to some degree with particles like electrons, but to think of bigger particles such as protons, much less big clumps of matter such as tables and chairs, is much more difficult.

We picture a proton as a fixed entity. In this theory, there are billions of Planck units twisted into a stable knot to form a quark. Quarks, in turn, are glued together in a very specific high-energy packet we know as the proton. The next level of complexity is that while the knot formation is extremely stable, the actual Planck units making it up, flow into and out of it in the briefest time. As if that were not complex enough, the individual Planck units become involved in the matter particle with an energy event similar to what we have been calling vibration in the electromagnetic wave. With the ebb and flow of Planck units through the particle with the intrinsic vibratory energy of that particle event, that energy is dispersed, to a small degree, to each neighboring Planck unit. It is the vibratory configuration of those Planck units in a specific pattern that defines a particle. The vibratory energy of each unit is actually spread out to its neighbors and also at the same time houses some of the vibratory energy of its neighbor.

The overall energy in a single proton is so overwhelmingly greater than the vibratory energy of a single Planck unit, given the billions of Planck units involved, that the uncertainty of the position of the proton seems less fuzzy and somewhat well defined, but at the quantum scale, it is not at all defined.

In the silliness of superposition-speak comes a useful concept: decoherence. Decoherence is the averaging of all of the stringy uncertain behavior of the individual Planck units as observed by someone looking at the whole picture of an entire quark or apple or planet. The uncertainty principle does exist in the macro world. The Planck units making up an apple share their stringy, temporal, imprecise vibration with all of their neighbors but from our observation point these imprecisions are washed out in the average behavior of an apple. Brian Greene describes this well: "[D]ecoherence forces much of the weirdness of quantum physics to 'leak' from large objects since, bit by bit, the quantum weirdness is carried away by the innumerable impinging particles of the environment."[57]

Light and matter travel through and utilize the same Planck units of the universe. There are no light Planck units versus particle Planck units. These units are ubiquitous. There is nothing special about any specific unit. A Planck unit thinks it is a photon one moment and a piece of coal the next. What happens when these two events coincide? When an electromagnetic impulse hits an object, the vibratory information defining the light impulse has to survive in a somewhat linear fashion as it winds its way through the intricate folds that make up the particle to emerge the other side as a recognizable light. That is to say, if a light wave strikes a lump of coal, it has to twist and turn through the Calabi-Yau manifold of quarks, intermingling with the vibratory fuzziness of those quarks and come out the other side without losing its characteristic light wave vibration. Obviously it does not do this. Almost immediately the vibratory information encoded as light transmission gets lost in the much higher complexity particle. This is the essence of absorption.

In order to pass through a solid object, that object needs to have enough holes that some light passes without ever getting caught up in the quarks. The latticework of glass, for instance, could be an example of this.

In summary, matter is the outward visible expression of one specific complex behavior of the Planck fabric of the universe. It took enormous amounts of energy to create the original package pattern, but the constituent units flow through these patterns continuously and do not claim permanent residency. Mass is a measurement of one of the expressions of matter as it finds its way through this all-pervasive field, suggestive of but not identical to resistance.

# Chapter 9

## FIELDS

A field is a very generic term. Fields can be used to define most things in our physical world. It is the expanse of space and time. It can include the entire universe or the space around an electron. The temperature and barometric pressure in your room are types of fields. The crystal orientation in the ice cube in your drink is a field. A field is a description of a property of a medium. It is not the medium itself but a characteristic of that medium.

We can measure, calculate, predict, and even manipulate field behavior using wave and quantum equations, but unfortunately this usually does not give us insight into the constitution of the media itself. If you were to picture a field as a sheet of dough and picture a particle as the donut cut out of that sheet, you could certainly look at the donut and get a clear understanding of what makes up the donut, or you could just as easily look back at the missing part of the remaining sheet and possibly come up with as clear an understanding.

While we may know exactly how much flour, eggs, water, and salt went into the making of our donut field, we have a less clear sense of what makes up particle fields. We know we can calculate them. We know how to use them to predict the behaviors of known particles. We even have been able to make field calculations to predict the existence of other exotic particles before we even isolated them. Much of the standard model owes its existence to such a thing. These predictions allow us to guess how to demonstrate new particles in accelerators.

Field theory is a strong, vibrant discourse in particle physics. We experience fields in our everyday life such as the gravitational field that sticks us firmly to the Earth or the magnetic fields that are so totally fascinating to the young child. As we grow up in this physical world, we learn that fields are a part of everything, so much so that we can describe everything by its field and field interactions.

The models that we build and the mathematics that we apply are logical and reproducible, giving them a high level of validity. Too bad they are based on a major misconception.

Consider a jar full of quarks (confined, of course) giving off their electromagnetic fields, gravitational fields, and local particle fields, stuck together with force-carrying gluons (force fields). The standard model suggests that associated with the nongravitational fields will also be their "messenger particles": photons for the electromagnetic force, gluons for the strong force, and weak gauge bosons for the weak force. So to continue the jar analogy, the three nongravitational fields give us three particles, which all communicate their messages via different languages to somehow communicate with one another and result in the behavior of the physical system within the jar. Note that the standard model, which evolved from relativistic quantum field theory, does not include the long-sought-after graviton for the gravitational force. The reason for this is that, prior to the advent of String Theory, general relativity's notion of gravity and its smooth space-time geometry does not work with quantum mechanics at the ultra-microscopic level, and thus the quantum field theory that works for the other three forces cannot be applied to gravity. So we have one set of rules governing the fields and particles of the nongravitational forces and another set of rules for the gravitational force.

On top of all of this, the Higgs boson model suggests that this complicated array of particles and fields interacts with Higgs field in order to impart mass. It's not exactly an elegant model, regardless of how frequently it is felt to have been experimentally validated.

The electromagnetic field is considered to be related to the wave/photon motion. The electric portion can be measured in the direction of the event and the magnetic portion seems to be perpendicular to that motion. We have already seen an alternative to the wave/photon duality model as a light impulse moves through contiguous Planck units. As that local motion is translated unit to unit, a ghost of forward motion is created and can be measured as the speed of light, though actually, nothing is moving.

This "ghost" motion, much like an ocean wave, has measurable properties of speed, amplitude, and wavelength. Using the model of an ocean wave, we picture that wave moving toward shore but also remember there is another motion, the up-and-down oscillations. The up and down deserves further consideration. As we know, the actual water molecules are not moving toward the shore, but they are moving up and down. This up-and-down motion is real and usable.

I will use my ocean turbine model again. Picture a contraption just off shore that is firmly fixed to the ocean floor. It has a float that moves up and down in response to the surface conditions so that as a wave passes, it will move through this cycle. This contraption could potentially be connected to a turbine so that the energy of the wave motion could be

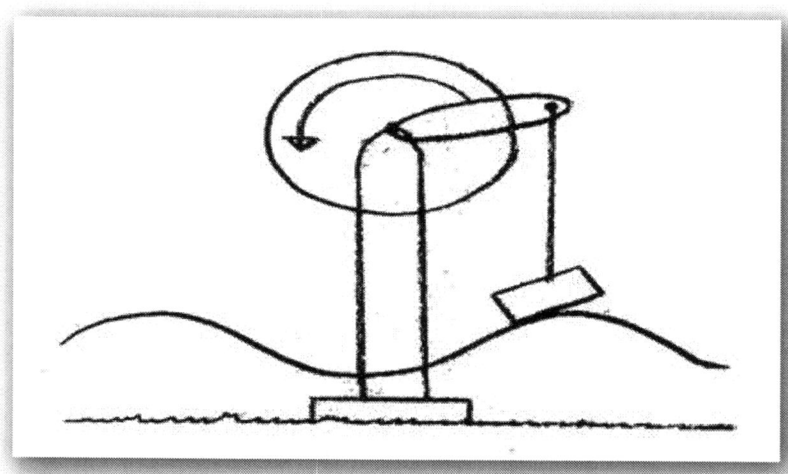

*Figure 5: The ocean turbine model.*

harvested. The forces applied to this float are not in the direction of the wave but instead are perpendicular.

This ocean float model gives us insight into the perpendicular "force fields" associated with wave motion. All waves have this perpendicular component. An obvious comparison can be made with the electric wave and its perpendicular magnetic force. A not-so-obvious couplet is matter and its perpendicular force of gravity.

You can see from the picture that after a complete up-and-down cycle, any specific water molecule moves back to its original position. The total sum of the vectors of motion over the course of one wave cycle is zero. Though no water molecules have actually moved anywhere, the moment-to-moment changes of the local up and down motion are measurable and usable. There is no sum motion of water molecules as a wave seems to move longitudinally toward the shore and no sum motion of molecules up and down as it completes a full cycle.

These force-field behaviors are a manifestation of ocean waves. They are a manifestation of all waves. They are a manifestation of waves as they pass through the quantum ocean.

It is clear that particle fields exist. They are exact, usable, and reproducible. Even so, they are still a little bit magical. There are no eggs, flour, and water to measure. The components seem componentless. Maybe we could rename our study "Faith-Based Field Theory," as we must believe it to be there without any true understanding of its component parts.

In particle physics, we can define the parameters of fields around various particles. These particles and their surrounding fields are quite exact and predictable, so much so that we can actually interchange particle for field in our equations and get the same result. This comes in handy, as some equations in particle behavior can be quite complex to nearly impossible but suddenly become manageable through the back door of field theory.

Picture a new, simpler model where a single quark is a superfolded mesh of Planck units. Not only is it superfolded, but these packages are moving through Planck units at incredible speed. The relatively stationary Planck units are enveloped into the quark as it moves through this field. A Planck unit takes up its position in this quark but only for the briefest moment, after which it is in its next position in the folded quark space. This process continues over and over instantaneously (one Planck tick) as each ubiquitous neighboring Planck unit makes its way, station to station, through the particle.

As we see in relativity, the Planck unit and the particle have equal claim on moving, one through the other. The quark moving through the field of Planck units is sucking these units up into and through the superfolds in a continuous fashion. This is how it is with all quarks, and everything else that quarks build, from a lowly proton to the largest galaxy.

This folding and unfolding of Planck units at light speed in space-time creates a push and pull on these units at the local level. The push and pull of billions of units folded into a single quark at any given instant imparts what we measure as mass. The push and pull is measurable as the particle's force field.

The perpendicular force resulting from this push and pull is a hole that we measure as gravity.

There are a limited number of Planck units that fill the universe. That number is unimaginably large, but even so it is finite, and that's all there is. So when billions of Planck units get sucked up into a quark fold, there is a relative local paucity. In the scheme of things, this paucity, for this one quark is so minor as to be negligible. A sum of an Earth's worth of quarks, however, is very noticeable. The sum of all these negligibly sized quarks makes one big Earth. The sum of negligible paucities adds up to one big, measurable void. The void surrounding the Planck units folding into and out of the realm of countless quarks produces a gradient. Just as air molecules move from high pressure to low pressure in what we sense as wind, there is an attractive force from the paucity of Planck units. This force is most pronounced locally and decreases the further one gets from the mass. It decreases by the inverse of the square of the distance. This force does not come from some magical and, as of yet, undiscovered graviton particle, but instead is one of the many manifestations of the one and only field, the Planck unit field.

The quark is a twisted knot of this field, a twisted knot of Planck units. It does not have a special, unique particle field. The fields are the expected effects associated with the twists and turns of these units in the multiple dimensions. There is no unique quark gravitational field. There is only the expected generic perpendicular pull from the local paucity of units as they are drawn up into these multidimensional patterns. This is how gravity magically acts at a distance from the superfolds of mass. It acts through the nearby Planck units, which are all interconnected.

This model of gravity is similar to general relativity's model, but with key differences. Both models hold that the effect of gravity is created when the fabric of the universe is warped. However, general relativity holds that mass acts on space-time and vice versa, but that they are still made of separate "raw materials"—that is, mass and space-time are different entities, but they are somehow able to act on each other. The Planck unit model, however, sees no separation at the fundamental level between mass and space-time. In this model, the fabric is all the same: mass does not somehow magically tell space-time how to act, but rather, mass and space-time act together because they are both part of a seamlessly connected mesh of the same material, Planck units.

It is supposed that every particle has its field and every field describes a particle. After all, field theory is a predictable, calculable study that can be used successfully to describe all of particle physics from electromagnetism to the boson. The Planck unit field is an idea that would help answer lots of difficult questions. If that field were to exist, then its specific measured disturbance could predict its specifically associated particle.

Please forgive my redundancy here, but I feel that I need to clearly lay out the foundation for each of these parameters of the nature of the Planck structure of the universe.

A theory about a field was developed not only by Higgs but also by Englert, Gamow, Alpher, Brount, Nambu, and

Goldstone—but nonetheless bears the name of Peter Higgs. The Scottish physicist's incarnation proposal basically added the Nambu/Goldstone boson, which removed inconsistencies (infinities) from the equations of previous theories.

This theory predicts the existence of a field matrix that permeates all space, everywhere, even in a vacuum, even deepest space. Its existence impedes other particles in motion. It is this impedance that imparts the impression of mass on matter. In this model, a photon moving through the Higgs ocean is not impeded and so is massless. A proton has a large impedance and so a large mass. This impedance is especially demonstrated when a massive particle has a velocity vector aligned with the force of inertia. The Higgs boson's existence was thrust upon us with no empirical data. Hopefully, one day soon, it will be taught as a historic attempt at the truth, much like the geocentric universe. It was assumed. It has enjoyed a long theoretical life. When searching for it, various sizes were estimated from 70 GeV to 1000 GeV. It was searched for but not found at all available energies of all available accelerators. In 2011, an event was noted at CERN. It should be made clear that in that much-celebrated event, there was no particle that was put on display for further evaluation. There is not a jar of these Higgs bosons sitting at the border of France and Switzerland.

It is clear that the detectors captured something, and the enormous computers interpreted that something. What that something was is really not so clear. It was a series of reactions that represent one potential theory of possible breakdown debris of one possible Higgs particle size. The event was captured as a triggered event. What this means is that of the almost infinite patterns of proton-smashing debris, specific patterns would tell the massive computers to take a look or ignore data. This technique is probably necessary due to the unfathomable amount of information generated, but it may introduce a significant, if not fatal, sampling error.

Triggering means discarding massive amounts of information. It captures about one or two ten-thousandths of the information created in the events. The rest is called background noise and involves known particles that are not under investigation, and so are excluded. Some wonder if this missing information could have been insightful (although not really since there isn't an actual particle to find). I would imagine if the general public were asked to explain their understanding of the Higgs discovery, it would go something like this: a proton smashed into another proton, and a new particle was formed. That particle was taken to the lab and investigated and found to exhibit properties that correspond to a field that imparts mass on all other matter. This is far from the truth.

There was a "something" detected at around 126 GeV, which is consistent with some of the predictions of a suspected Higgs entity. It was measured as various different decay patterns in various channels and additively these findings reached a critical probability level called 5 sigma, or five standard deviations from the mean. This corresponds to a likelihood of a one-in-a-million chance of occurring by chance and therefore not likely due to chance: a statistical way of saying, "I'm almost positive this is the truth."

Suppose we had a theory that a peanut weighs one to five grams. We then went to the state fair and found a game in the midway where we throw baseballs at plates. Plates are broken each night. Unfortunately, we could find no pieces weighing one, two, three, four, or five grams. At this point we arbitrarily decide that a peanut weighs 1.26 grams. Thousands of plates are broken over a particularly busy night. We collected all of the shrapnel of plates and measured the weight. We find one fragment that weighs 1.26 grams, which fits with our original theory of peanut, so by definition this must be a peanut. Or at least it is within five standard deviations of being a peanut. But wait, we didn't actually get to hold the 1.26-gram piece in our hands, because immediately after collection, we ground up the bits into sub-bits. Fortunately, we know that out of the 250 pounds of debris, these certain grains of ground up plate bit could only have originally come from a 1.26 gram piece that must have been the long-sought-after peanut. We would gladly pay up to seven billion dollars for a carnival game that could provide that kind of information.

The initial particle itself was never seen, as it decayed within $10^{-22}$ seconds, not even enough time to clear the radius of the proton it came from—certainly not enough time to reach a detector and in no way was visualized as its own entity. Even the presumed breakdown products lasted only 1/1,000,000th of a second.

Isn't it odd that a boson defining an all-pervasive, active, present field that imparts mass to all matter and exists in every nook and cranny of the universe cannot be found in nature and when "created" is so unstable as to decay in 0.0000000000000000000001 seconds? Obviously, with so brief a visit, no Higgs field was measured in this historic event. Shouldn't the particle be as prevalent as the field it defines? Shouldn't it be an especially stable ubiquitous particle to surround itself with a field that essentially defines all mass?

We assume that this particle exists because it is celebrated and taught and hoped for and fits with our theories, but this one makes no sense. This is an example similar to the parable told by *The Emperor's New Clothes*. All of the people, deep in their souls, believed the emperor was wearing beautiful new clothing. No one even recognized that he was naked until some naïve child said something, and suddenly everyone recognized the truth. Few voices have been willing to suggest that these findings make no sense. That level of critique would necessarily have to come from the physics community. The rest of the world has no access to particle physics or the high-level math behind it, and so all of this is assumed to exist as hard fact. There is virtually no critical oversight to this process, and there are vast sums of money at stake, and so here we are with conflicting purposes. We have a field that imparts mass to all particles and that fits well with Dr. Higgs (et al.)'s equations, but no available particle. We need this particle because without it, we are left with the conundrum of a missing explanation for "all fields define a particle, and all particles have a field."

Aha, the key! What if the particle does not own its field? What if, instead, particles interact with a general universal field? What if there is no difference between particles and fields, but they are instead manifestations of the same physical entity?

There is a field that is spoken of by various names, depending on the specific behavior under review. Every time mention is made of the existence of an all-pervasive something, it is this field. It sits there, undetected unless disturbed by some local event, a proton passing through for instance. The proton would interact in a predictable pattern so long as each proton is precisely the same everywhere in the universe. That would further suggest that the field is precisely the same and uniform everywhere throughout the universe.

We see that gravity and magnetism act at a distance. Here's a shocker: photons and even mass act at a distance. They do so by means of their fields. Everything has a field and yet we are hard pressed to really get our hands on one. That is because **there is one and only one all-pervasive field!** This quantum field, then, is everything. It interacts with everything. It defines all of electromagnetic behavior, all massive particles, even gravity. It stands alone. It does not define a single Higgs boson. It describes all events and all particles from photon to universe.

All particles have a field: the Planck quantum field.

This field has its particles: any and all particles.

Higgs ocean may have been a reasonable starting point. It may hold scientific value and acceptance because it is predicted by the complex equations of quantum physics. In this model, Higgs field is a medium through which other things move. This just barely scratches the surface of what this field represents. And yet, somehow, no one has noticed that it totally replicates the theory of aether.

# *Chapter 10*

## PRINCIPIA (GRAVITY)

Gravity is pretty darn important. Sir Martin Rees's words on the weakness and grandeur of gravity in his book *Just Six Numbers* are beautifully descriptive and particularly insightful. Prior to Newton, objects fell to the ground because it was down. Obviously, things fall down; everyone knows that. But suddenly a new paradigm arises that says no, there is a primal "force" at work here that can be absolutely defined. Not only can we explain a falling apple, we can explain even the behavior of the planets orbiting the Sun and the moon orbiting the Earth.

Despite these mind-boggling feats, gravity is still "amazingly feeble."[58] It is by far the weakest of the four classic forces: strong, weak, electromagnetic, and gravitational. We feel the profound effects of gravity when jumping into the air and quickly coming down, but the force of gravity is only approximately $10^{-36}$ times that of the electromagnetic force. Consequently, it plays a very small role in the particle world. Compared to the other atomic forces, gravitational effects are negligible and are easily overwhelmed.

As we move away from particles and into more massive objects, the effects of gravity increase. Gravity increases proportionally to the mass of an object, and that being the case, gravity has a profound effect at the macroscopic scale. We are familiar with many of these effects. Gravity makes things fall to the ground, keeps us stuck to our planet, and in fact is what keeps our moon in its orbit and the planets safely in our solar system. That the gravitational force is dependent on the mass of an object as well as the distance between objects has been with us since Newton's time. It obeys the inverse square law first described by Sir Isaac Newton in the *Principia* in 1687.

As background, let's see some of the neat tricks gravity has to offer. Fortunately for us, it is only as powerful as it is. Any stronger or any weaker, and we would not be here to consider it.

Asteroids of a low mass (low compared to planets) do not exert enough internal gravitational pull to overcome the friction of their own particulate matter and so retain their irregular shape. Each speck of asteroid dust certainly feels the gravitational pull toward the center, but that pull is too weak to shift these particles significantly. The asteroid therefore gets to maintain its irregular character rather than being pulled and molded into its lowest energy shape, the sphere. If gravity were a little stronger, asteroids would become spherical. Stronger still, and objects of even lower and lower mass would get crushed into a global shape. Only a couple of orders of magnitude stronger (but still $10^{-28}$ that of the electromagnetic force), and ordinary earthly objects would crush into globes. No more chairs, tables, cars, trees, people. There would just be lots of spheres.

Obviously, under those conditions, life would not have been possible. This wouldn't matter too much, because at that level, the gravitational force of the Earth itself would be so powerful that it would crush into itself, forming a small, very dense black hole. That event, too, wouldn't stand a chance, because gravity would have pulled the Earth into the Sun way before it had a chance to form. That wouldn't matter very much, because our solar system and the rest of our galaxy would have been sucked into Sagittarius A, the black hole that presently supports the Milky Way. In this case, feeble is very good.

Our universe's value for gravity has allowed everything to unfold. As opposed to asteroids, once matter reaches planetary mass, it is pulled into its somewhat spherical form by its own gravity. As the mass increases, the self-gravity becomes stronger until it becomes so great the planet crushes itself. Planetation can only support a certain size, the maximum being approximately that of Jupiter. Any chunk of matter greater than Jupiter will be under so much pressure that it will collapse into a black hole, unless it has an equal outward pressure such as occurs in the nuclear furnaces of stars. According to Rees, if our Sun were cold, it would become a white dwarf. It would be millions of times denser than ordinary matter.

It would be approximately the size of the Earth but 330,000 times more massive.

In 1915, Einstein's work on general relativity showed that gravity and accelerated motion are indistinguishable, and this allowed him to ultimately produce the equations that showed that gravity is the warping of time and space. In 1916, a German astronomer, Karl Schwarzschild, used Einstein's new equations to make some mathematical predictions of his own. In particular, he showed that if a star were dense enough so that its mass divided by its radius reached a certain value, it would cause an intense warp of space-time. Gravity would be insurmountable, and everything within its critical radius would be unable to escape. This gravitational field is so intense, so complete, as to not even allow something as evanescent as light to escape. This was later called a black hole, a term coined by John Wheeler in a talk given at the NASA Goddard Institute of Space Studies in 1967.

Black holes are intense concentrations of mass. They have reached that critical density and they collapse on themselves. By definition, the density of this mass is so great as to have a gravitational field that will not allow light to escape at any cost.

The study of black holes, although mostly theoretical, has been very insightful into the behavior of gravity. The study of black holes is not actually the study of the nugget in the middle or even a study of the holes, but rather a study of the behavior at the event horizon. Black holes are not some mysterious entity but an expected consequence of extremely large collections of matter without enough kinetic energy to offset gravity. That being the case, we get to see gravitation at its most extreme.

The simple mechanism of gravity has been one of the more stubborn aspects of physics. We definitely understand *how* it behaves, just not *why*. Newton helped give us a user's manual—but without explaining the causative etiology. Einstein let us in on the secret that it was not an actual force but a result of the bending of space-time. One way we know gravity is not a real force is that it is only attractive. It is not two-way, as any force equation would dictate.

Quantum theory, string theory, quantum loop gravity, and their variations take stabs at the nature of gravity, but all wait for the future to explain the mechanism. They all believe it is just around the corner, but as yet, the actual action of gravity on another object at a distance without direct contact is not fully explained.

The concept of a distinct gravity mediating particle, the graviton, is an absolutely useless grasping at straws. It is assumed that every field has a corresponding particle and gravity is a field if ever there was one, so by default it needs its particle. It depends what your definition of "is" is. If we are willing to expand our definition of what "particle" could be mediating this process, then maybe in a loose sense, that particle exists.

Even without the graviton, Einstein's general relativity seems to lack a completeness that explains the relationship between space-time, mass, and gravity. The key conclusion of Einstein's idea of gravity is that mass somehow acts to warp space-time. General relativity holds that mass and space are connected through gravitational effects; it does not account for *how* mass acts on space-time, how these two entities speak to each other. Einstein's model maintains that mass and space-time are still different entities, regardless of how intertwined they may be. But perhaps we can take Einstein's vision of gravity and space-time one step further by imagining mass and space-time are able to seamlessly communicate and affect each other because they are manifestations of the same entity.

As you should have surmised by now, Planck units are absolutely everything. There is not an object or event that at its most basic level some behavior of Planck units. Planck units, however, are ubiquitous and unremarkable. Each unit is identical to the next. Each unit is surrounded by its neighbors. Planck units have no self-awareness. They sit quietly until asked to transmit information but without any sense of purpose. As far as they are concerned, they just exist. Little vibrations happen now and again but basically they go through eternity unchanged. They don't have any knowledge of whether they are sitting in deep cold outer space or part of a photonic vibration or folded up into an atom. There aren't separate "gravity" Planck units, "mass" Planck units, "electricity" Planck units, and "magnetic" Planck units.

Planck units are the mediator of gravity along with every other thing and event. **That ubiquitous Planck unit is the graviton.**

Picture a cube of space of quiescent Planck units. For simplicity of this mental model, we will have them all aligned in rows. As a random atom passes through this space, it incorporates these Planck units into, through, and out of itself. Not only is the individual unit unchanged by the experience, but it remains in direct contact with its neighbors throughout the process. It has no sense that it has been pulled briefly into and through the atom.

As matter is formed, Planck units are pulled up into its various folds. In this process, the perception and behavior of the surrounding milieu is altered. As the fabric of the cosmos is gathered up into its convoluted multiple dimensions, its local environment changes in some manner.

If you were to step from one Planck to the next, it would seem that you are walking in the same straight line, even though from afar it now looks like the entire area has been twisted. The warp is not in the units themselves but in spacetime: it is the geodesic, a straight line in curved space.

Gravity is the consequence of and also defining force of all of astrophysics and actually particle physics as well. Quantum loop gravity has all the necessary ingredients for a complete theory of everything if its proponents were not so stubborn and even condescending about the aether concept. Some even describe the atomization of the universe in Planck-size quantum steps as they belittle an all-encompassing milieu.[59]

It is common to use a two-dimensional model for this three-dimensional process. We build a model with a membrane stretched by a heavy object—a trampoline with a bowling ball, for instance. Just as the ball makes a deep depression in the trampoline, the mass-gathering process pulls the fabric of the universe locally into its convolutions. This affects the fabric area immediately surrounding that mass unit. This creates a geodesic impression (the bowling ball representing Gaea). The indentation that this ball makes encourages other matter to fall in toward it at an acceleration that is dependent on the ball's mass. This is similar to the property (or impression) of gravity.

Whether the Planck units are actually stretched under the tension of nearby matter or matter merely changes how we experience the surrounding area depends on our viewpoint. The sense of gravitational space-time bending is a matter of perception. That is to say, if we are viewing the space-time bend from a distance we can perceive the bend, for instance light bending around a distant star. If we are part of the bent space, then we are bent along with it and all seems linear again. In any case, we thankfully continue to stick to the ground.

It must be remembered that this planar model is vastly oversimplified, as these geodesic paths are actually within the ocean of Planck units in a three-dimensional space going in every which way; we are not dealing with just the flat sheet of a trampoline mat.

It may be difficult to conceptualize how this apparent distortion in linear motion can translate into an object being drawn toward another. It is because everything is moving through space-time at the same rate, the speed of light. Light moves through space at 186,000 miles per hour, so in space-time its movement is entirely in the space component. We move through time at much closer to zero miles per hour, so our motion through space-time is measured more in the time aspect. Everything moves through space-time at the same rate, which is a combined function of how we see each component separately. As the Earth and our feet move through space-time together, the mass forming process of each element warps our experience so that we are sucked through it together. Relativity tells us that since information travels at the speed of light and since we are so close to the Earth, for our limitations, we are essentially the same spot in the universe as the Earth.

We can't really say what our velocity vector is in reference to the grand framework of the universe. As Einstein taught us, motion is a nonsensical notion outside of a frame of reference. Even so, as a fun exercise, assuming that for a moment all of the following vectors of motion are in the same direction, consider that if you are walking on the equator at three miles per hour, the surface of the Earth is moving at approximately one thousand miles per hour. We revolve around the Sun at approximately sixty-seven thousand miles per hour. The Milky Way is a flat disk eighty thousand light-years in diameter and some six thousand light-years in thickness with a spherical halo of stars extending out for about a diameter of one hundred thousand light-years. Embedded in this disk is our solar system, at about thirty thousand light-years from the center, just off plane. The disk rotates with a linear speed at our solar system of about 490,000 miles per hour. Based on measurements from the COBE satellite, the Milky Way is moving at about 870,000 miles per hour in respect to the cosmic background radiation. Based on this speculation and with the aforementioned assumptions, we are moving through space at almost 1.5 million miles per hour (speed of light 671 million miles per hour).[60]

If that were the case, our space component in the space-time context would be 0.002 percent that of light, and our time component would account for the remainder.

A curve in space-time appears the same as acceleration. An angular change in momentum is measurable as a force. Gravity gives an impression of force. For most everyday purposes, it follows the Newtonian rules of force and can be calculated in that fashion. For the surface of the Earth, we feel a force equivalent of 9.8 $m/s^2$.

General relativity tells us that gravity is not an actual force but a bending of the space-time continuum. That light that seems to bend around a distant star is, in actuality, following a straight path through a curved space-time, its geodesic path.

Gravity is to mass as magnetism is to electric impulse. It is a stringy behavior. As an example, an electric impulse travels through its Planck unit pathway, causing a vibratory disturbance. The local compaction and rarefaction of these units results in a perpendicular force exerted on its own Planck pathway. This perpendicular force we call magnetism.

It is through these changes that occur to these local but perpendicular Plank units that magnets appear to act at a distance. In a similar, but maybe more complex manner, gravity acts at a distance. A knot of units making up matter moving through the universe enveloping new Planck units while leaving the old behind. This causes the impression of mass in our three-dimensional observation in a similar stringy manner. Perpendicular to the bulges and valleys of disturbed units that we call mass, another stringy process of reciprocal disturbance is caused. The Planck units "send out" this disturbance, with an attractive force proportional to the mass. It is through these altered, perpendicular Planck units the force field called gravity appears to act at a distance.

It takes significant mass for us to feel gravity, but the essence of its mechanism is at the minuscule Planck level. Our gravity is the summation of all the Planck space curvature brought on by all the mass of the universe diminished by the square of the distance any speck of that mass is to us. While this force may act at a distance, it dissipates quickly according to the inverse square law. It gets negligible after a short distance but, at least in theory, acts out to the far reaches of time and space, approaching but never reaching zero. The effects of gravity are additive, with all mass in the universe adding its contribution.

Though this mechanism, the graviton is not its own particle but rather is the gravitational behavior of the Planck unit. It is in this manner that the Planck unit is the equivalent of the graviton, and there is no other need for a fictional particle to be created. It is the graviton only in that briefest time that it is participating in any gravitational process. A moment later that same unit may be swept up into the folds of a quark and then, a moment later, vibrate in an electromagnetic impulse.

Mass and gravity are not primary entities. They are properties of matter. They are measurable manifestations of Planck units. Matter gives the quality of, or the impression of, mass and gravity. Planck units themselves do not have mass and so do not bend space-time and do not display gravity. Quiescent Planck units in deep, dark outer space are the extreme of nonspace-time bending. When we think of matter, we see the other extreme of space, the space-time warp of general relativity. That is to say, a large body of matter with large mass bends space-time as much as it can be bent. Black holes bend space-time extremely, so we have it in our minds that this extreme display of mass consists of very, very dense matter, and yet the universe might have still another twist.

Our galaxy is held together by the gravity of one of these massive black-hole structures that has been named Sagittarius A. It has the mass of 2.5 million suns and a diameter of 12 million kilometers. These obviously are not measurements, as a black hole cannot be seen and measured, but are estimates based on the gravitational behavior of the Milky Way.

As big as that is, astronomers believe there are black holes out there with mass as great as several billion suns and a diameter comparable to our entire galaxy. A problem arises however in the maintenance of the shape of our galaxy. We are used to the model of our solar system where the gravity of the Sun holds our Earth in orbit. On a galactic scale, the stars at the outer limits will be moving much faster through space while also sensing less of the gravitational pull of a central black hole to the extent of the level of the inverse square law. If the entire galaxy is held together solely by the mass of Sagittarius A, then the furthest stars should fling off into space. Even with the additional mass of all the stars and their planets, there is not enough mass to account for our outer stars maintaining orbit.

This would suggest an enormous amount of unseen force scattered evenly throughout the galaxy. The gravity-forming entity may be numerous black holes or dark matter or dark energy.

It is possible that black holes could be small. Small, scattered black holes, even at an unfathomably small size, would still have extreme density of mass. To get an idea of that density, in order for the mass of our Sun to reach black-hole density, it would have to be compressed to about three and a half miles in diameter.

We get this compression picture in our minds, and it makes us think that a black hole is the ordinary matter of this world with all the air taken out of it. There is no space between Planck units, and so a black hole must be a special set of circumstances where a critical mass has caused an extreme level of folding so that an enormous amount of Planck units get twisted into a small space. This may be the character of a singularity.

Gravitational attraction is experienced by all participants with mass. In a system with two bodies in study, each body will experience the gravitational effects of all of the mass of the system. The bodies in question do not feel the pressure evenly, however. Since gravitational force decreases by the inverse of the distance squared, the nearer side of an object will feel the effect more strongly than the far side. This explains the high tide of the side of the Earth nearest the moon. It pulls the remainder of the Earth so that the sides of the Earth are pulled in, or low tide, and the Earth itself is pulled away from the water at the opposite side from the moon, creating another (lower) high tide.

# *Chapter 11*

## TRUTH AND BEAUTY

The artist sees it; the pessimist doesn't. There is beauty in everything around us. The universe bathes us in its splendor.

It seems to be human nature to seek out the answers to life's ultimate questions, such as the origin of the world and its components. Some of us approach this through spiritual means while others tend to look in a science book. Even the most religious of us recognize that science has a role in this description. There certainly is a beauty in the theories and equations in the latter approach. Complex events can usually be broken down into their various component parts through the intricate considerations of equations. Those who see beauty in math often recognize this beauty in its symmetry.

Symmetry can mean many things. Symmetry is the circumstance when an object remains unchanged through various conditions and orientations. Symmetry is all around us. It is natural. That is, symmetry is a phenomenon seen repeatedly throughout nature. Physics and math often help define it. We see symmetries throughout our day without even taking notice. It makes the world around us more easily understood and quantifiable.

In geometry, symmetry is seen in all sorts of shapes. The easiest symmetry to see is a ball, a sphere. Turn it any which way, and it looks the same. This is rotational symmetry. Move it to another location, and it still looks the same. That's translational symmetry. After a good night of sleep, you come back and see that thing still looks the same. That's temporal symmetry.

If everything looked like a sphere, the world would be easily quantifiable. Obviously, it is the nonsymmetrical events that give the world character. Even a sphere can "break" symmetry. Put a ball in a perfectly square box. It will rest on the bottom of the box. This changes the nicely symmetrical box to one that has a nonsymmetrical blemish. This is a visual representation of symmetry pervasive throughout physics literature, but it is only a model of the true symmetry in question.

Symmetry in physics is actually an arithmetic event. It has mainly to do with how events behave in time and space. The math of symmetry study is usually expressed in group theory. Group theory uses a table or matrix to list the items. It has its own unique jargon that infiltrates particle physics literature.

The evolution of particle physics, unfortunately, has traveled down the wrong path from its inception. Each wrong theory that we cling to so dearly requires mathematical gymnastics to keep alive. And so we are left with many attempts at explaining a treasure trove of inconsistencies. The many fudge factors formulated to renormalize these inconsistencies, some of which are very complex, some of which are very abstract, appear to be with us to stay. They really either need to be explained away or enveloped into the actual way the universe works. In the exploration of symmetry, we can regain some of our footing.

Though it may be asymmetry that imparts character to our world, it is the role of science to break the big stuff up into its component parts until symmetry emerges. We separate, define, categorize, and distinguish. We study things and list them and then break them down even further until we reach a point of atomization where they are at their most essential level of existence. Here we find symmetry.

So what does this have to do with symmetry in the ultimate truths of the universe? Once objects and actions are broken down to their most elemental level they can be compared for sameness. This sameness can be considered a truth, a proof. Now that it is established as a truth, if it were moved to a new unique context, we could either predict how it should behave or, alternatively, observe how it actually behaves and gain insight into the system it was placed.

At Planck's length, objects and events (time) are at their most elementary, atomized level, and this establishes true identity and true symmetry. It is only through breaking everything down to this essential level that we are able to start building everything back up in a sensible way. All the complex asymmetries that give character to our surroundings are a buildup of Planck units, wherein lies the symmetry.

The most perfect symmetry would be an undisturbed field. Picture a giant swath of outer space in a prehistoric moment when there was no mass or gravity to apply their effects. Every possible measured parameter in every possible orientation would be completely uniform. An immense ocean of untouched Planck units. An immense ocean of boring, motionless, uniform-temperature, minuscule dots.

As beautiful as this extreme symmetry may be, we would not exist if the universe had been born to this level of integrity, an amorphous uniform universe made up of the most basic units. Fortunately for us, through chance or temperature variations or some sort of cosmic Brownian motion, glitches did occur. By chance, in the great expanse of everything, there would be moments of concentration and areas of void, compaction and rarefaction, ripples and antiripples in the quantum foam[61] or quantum jitters.[62] In "Higgs ocean," small amounts of whatever is there form up and decompose constantly and randomly. A sort of weather system develops with vast expenses of these variations. An extreme web of energy fluctuations moves from a quiescent lowest energy state to ripples and antiripples. Movements fold and unfold, and sometimes—rarely, but sometimes—a magical moment occurs where they amplify until they attain a stable configuration: a particle.

In theories solidified by German mathematician Amalie Emmy Noether, quantities such as linear momentum are conserved no matter what physical circumstances they are placed in. These quantities, which are not dependent on place and time, are said to be invariant. Her theories were of universal scale and said to be "global" (I guess because our globe is the only part of the universe that matters).

Another German mathematician, Hermann Weyl, known to Einstein and fascinated by Einstein's general relativity, applied Noether's invariance theories to local events. Rather than studying the effects on quantities when the universal coordinates are changed, he applied his attention to the invariant nature at the local level, such as distance of train travel in space and time. He used a form of math listing continuously changing groups developed by Sophus Lie, a Norwegian mathematician, 1842-1899. As we look at the world around us, we see that the goings-on are very complex. Group theory helps codify things at their most basic level. It is an attempt to turn what seems outwardly random into a logical list for further investigation. This change is called transformation and is applied to the symmetry of these groups.

It is Weyl's use of the train model at steady velocity where the engine gauges are fixed at one value that led to the unfortunate adoption of the term gauge to mean local in today's physics parlance. He did, however, use local invariance applied to Einstein's general covariance to combine electromagnetism with gravity. Einstein did not approve, and so Weyl dropped the issue.

Erwin Schrödinger sensed a connection between Weyl's gauge theory and electromagnetic waves to develop his theory of wave mechanics. Using the terminology of Lie groups, U(1) symmetry refers to a matrix or grid of values under the condition of one complex variable. For instance, a circle is a model that we can easily picture as a continuous function. We can choose angles to stop at or measure, but the essence of the circle is continuous. Rotating this circle to any degree, as long as you don't change the two-dimensional plane it sits on, will not change its symmetry. The act of turning the circle is what is meant by transformation.

A sinus wave can be measured in a way identical to a circle. Moving along the wave is a function of distance and time and, with each step, follows a cyclic measurement. This movement is a transformation and is completely repetitive and predictable, just like a second hand going around a clock face. This repetitive nature means any and all measurements are repetitive as well. It is invariant. Schrödinger recognized this, which eventually lead him to the quantum electrodynamics of electron wave function with respect to invariant values for electric charge.

There is a story of superconductivity that led to some of the early thoughts on string theory. In 1957 John Bardeen, Leon Cooper, and John Schrieffer (known collectively as BCS) described some unexpected electron behavior. The simplest model of superconductivity is the conduction of electric charge. In normal electric conductivity, a force is required for the movement of electrons against a gradient. However, under certain circumstances (such as very low temperatures), the electrical resistance of certain metals will fall to zero (or nearly zero) so that electrons can glide freely. In this superconductor environment, application of a magnetic field results in a separation of field lines. These field lines represent the fundamental units of magnetic flux. These lines are not arbitrary. They fall into a predictable pattern and are always a multiple of a fundamental integer: a magnetic quantum, so to speak.

Thousands of pages of text, equations, and theories later, it was recognized that the binding forces between elementary nuclear particles "quarks" followed a similar behavior. There appear to be discrete force lines connecting these quark particles, the strong force, the gluons. These force lines exist in nonarbitrary values, integers of a fundamental value. These strings follow the superconductor behavior seen in BCS theory and represent the origins of string theory.

The force-carrying particle of electromagnetism is felt to be the photon, which is, as we now know, actually a specific behavior of wave transmission through the Planck field. The force-carrying particle of gravity is felt to be the graviton, which is a specific wave transmission through the Planck field. Weak force-carrying particles are some massive bosons, but I leave that for the next chapter. So the force-carrying particle for the strong force, the gluon, is the sum energy entrapped within the complex manifold of a quark or, in other words, a particular behavior of the Planck field under the special conditions of high energy.

Spontaneous symmetry breaking is the term used to describe when any stable symmetrical system undergoes a transformation, which obviously results in a change in the system. This would result in a loss of the original symmetry. It has pretty generic meaning but is a theory applied to particle physics. In this application, a stable symmetrical system goes from its original energy state to a different, lower energy state. In this process, force lines occur much like the field lines found in the superconductor experiment above. They are found to occur in discrete steps corresponding to quanta (Planck units). It is felt that these force lines are responsible for holding things together, quarks, for instance, and represent the strong force of particles. It is further felt that through this process of symmetry breaking and resultant strong bonds that the quality of mass is imparted on particles. Unfortunately, as with everything else in today's physics, we've collectively decided to devote some massless particle to mediate this process.

Yoichiro Nambu and Jeffery Goldstone started work on broken symmetries in superconductivity and recognized how this might account for mass of particles. The broken symmetry would change in a particle from massless to massive under the influence of an energy source that was hidden but percolating around in empty space. Equations initially seemed flawed, and it appeared that his theory would result in some unwanted and unaccounted for massless particle.

Symmetry breaking is also how we move from the electromagnetic "photon" behavior of Planck units to matter's "mass" behavior. That distinction may not have been evident in the earliest moments of the universe, as the temperature was too high for anything more than particle plasma soup. To arrive at symmetry breaking, we need a starting point of unbroken symmetry. This would the pre-Bang state of the universe, before there was a singularity. Before anything mass-like. In Nambu's model, this could be considered the initial state of unbroken symmetry.

In the setting of the superhot high-energy early universe, the knots and twists that represented prematter could have been in a fluid, hyperexcited state called plasma. This plasma state is very reminiscent of the superconductor model for electron movement. In this model, we could consider this initial high energy plasma as the original symmetrical state. Although it is hard to call anything with that much energy around spontaneous, it is the falling from one energy state to a lower state that transforms that system.

In this system change, some of this available kinetic energy is aligned in the form of force fields to induce bonding. This is the moment of subatomic matter production as well as the induction of the property of mass.

Peter Higgs embraced this particle, gave it mass, and promoted this theory. Others also worked on Nambu's theory, including Robert Brout, François Englert, Gerald Guralnik, C. R Hagen, and Tom Kibble. They all worked out various systems for how broken symmetry resulted in particles attaining mass. It turned that all theories were similar and could be converted one for another, which was felt to give support and credence to their theories. Partly by chance and timing and partly due to one particular addendum to Higgs's paper that added a newly envisioned, theoretical boson, the process and its boson are forever associated with Higgs.

We so desperately wanted to isolate a Higgs boson that we have spent billions of dollars to do so. As we've discussed previously, it was even reported in 2012 that it had been found, although this is not as clear-cut as it sounds. The Higgs boson has been estimated at various sizes, but by one estimate at least, it is 126 GeV. It is believed that at this size, it would break down in one of several predictable patterns. Then we smash protons together for several years, creating all manner of byproducts. When one of them by chance falls into this mass range (126 GeV) and decomposes as predicted, we say, "Eureka!"

Earlier, we used the peanut example to demonstrate the ridiculousness of the methodology that accounted for the "discovery" of the Higgs boson. Here's another analogy. If you threw ten coins, the chances of all of them landing on heads is one in 1,024. This single event would be four standard deviations away from the expectation. So if all came up heads, one could say that most likely, the coins do not land randomly. They must be more likely to come up heads. Finding the Higgs boson was like this. It was a 5 sigma event, meaning more than five standard deviations away from the expectation for a single event. This is what allows the claims that this discovery is real and not likely due to chance. The rules change, however, if you do the experiment one hundred million times. If you threw ten coins that many times, they would come up all heads 97,656 times. If you smash protons one hundred million times, sooner or later, you will create a particle of the predicted size.

Plenty of—almost infinite—unstable particles formed and disintegrated immediately after the Bang. Every possible pattern of knots and twists of every possible size were created and destroyed. Different sizes have been attributed the magical Higgs boson: 114.6 GeV, 160 GeV, and 270 GeV. Particles of all of these sizes were surely in great supply, but those sizes themselves don't infer any special properties. Furthermore, just creating such a particle in an accelerator does not assign it any special importance, either.

Though it could be reasonable to think of a Planck unit as a Higgs particle, it by no means is a boson and does not have mass. Furthermore, it is inexorably enmeshed in the blanket of what would otherwise be considered Higgs field. The meshwork is so inseparable that even the Big Bang could not tear it, much less could our little bangs from the accelerators. As much as we attempt to understand the Higgs postulate in terms similar to other primary particles and their fields, it is, as is everything else, a particular manifestation of the Planck field. However these units may exist, they cannot be separated from their neighbor or their position in the matrix in general. These are by definition the smallest thing. It is what Planck's constant describes and is most likely on the order of $10^{-35}$ meters.

A Higgs particle, as postulated, has mass. It is associated with the Higgs field, which gives mass. So this massive particle depends on its own existence to impart mass on itself? A Higgs boson will never actually be found, because it doesn't exist. The only "particle" that is associated with mass is the

Planck unit, and it will not be separated from the field. There will be no isolating this in an accelerator. There will be no tear in the fabric of the universe.

The idea that the Higgs field can be switched on and off is misleading as well. Although the actual mechanism of mass impression would be ever-present, at the earliest moments of the universe, the kinetic energy was too high to allow twists and knots to find stability and not be blown apart as soon as they form. As the universe matured and cooled, a temperature was reached that allowed this balance. It has been referred to as breaking of "electroweak" symmetry. The balance between high-enough temperatures and kinetic energy to create and low enough to keep from blasting apart was the breaking force not some previously unknown field or particle. There is no need for some massless hyperenergetic particles (tachyons) to condense, allowing some mediator bosons to attain mass. This condensation was due to cooling, which could be considered spontaneous and could supposedly have been the original symmetry break.

After the Big Bang, there came a point when the kinetic energy was sufficiently low to allow the Higgs field to be "turned on." What that says is that while prior to that moment there may have been sufficient energy to create knots of heavy nuclei, the kinetic energy and radiation were so high that any atoms created would necessarily be rapidly destroyed. We could even postulate the timing of the creation of certain essential particles based on their intrinsic energy, $E = mc^2$, and the available energy as the universe cooled. Larger particles (protons, neutrons) would have required higher energies and would have been twisted earlier than lighter electrons, but at those high energies, there was extremely high radiation counteracting the nucleosynthesis.

Symmetry would dictate that in the first moments of creation, for every piece of matter, a piece of antimatter would also be created. Maybe at superintense temperatures they could coexist, but soon after creation they would have the tendency to annihilate each other. Equal particles and antiparticles[63] would be very bad for us because we are made of this matter stuff, so there had to be leftovers. There had to be an excess of matter over antimatter, and that means we had to make up some explanation. It is estimated that there was an excess in the ratio of 1,000,000,001:1,000,000,000 matter over antimatter. After the annihilation, 0.000000001 of the original matter would have been left over to make up all of today's ordinary matter for the entirety of our universe.[64] There are explanations for this, involving moments soon after the Bang when spin parity, charge, and thermodynamic symmetry could have been violated, resulting in the mismatch. To say the least, this seems like an arbitrary and complex way for the universe to go about things.

Sometime early in our Physics 101 course, we are told that in the duality of nature, everything expresses a wave-form function, including particles. It is simple to imagine a light beam traveling in a wave fashion and a photon particle because the size and scale put them out of our ability to perceive them, but it is much more difficult to picture a baseball or car with a wavelength. In nature's symmetry, all electromagnetic and matter events would manifest both a wave behavior and a particle behavior. Louis deBroglie conjectured this wave behavior for an electron traveling in its Bohr's orbit around a proton, H.

A physicist at Western Electric (Bell Labs) named Clinton Davisson was experimenting with cathode ray tubes in 1927. While playing with these tubes, he unexpectedly demonstrated that electrons can display an interference pattern consistent with waves. The very foreign idea that a particle with measurable mass could behave like a wave, which, as mentioned, was first introduced by deBroglie only a few years earlier, was now experimentally established.[65]

Imagining that electrons exist as a wave function is still acceptable to us because they are too small to see, but how about the bigger stuff that fills our daily lives? Unfortunately, applying a wave concept to more massive objects has been very difficult to palate, even for our greatest thinkers. Brian Greene touches on this apparent dilemma in *The Elegant Universe*, asking the question, "But how does this jibe with our real-world experience of matter as being solid and sturdy, and in no way wavelike?" and offering the following explanation:

> Well, deBroglie set down a formula for the wavelength of matter waves, and it shows that the wavelength is proportional to Planck's constant *h*. (More precisely, the wavelength is given by *h* divided by the material body's momentum.) Since *h* is so small, the resulting wavelengths are similarly minuscule compared with

> everyday scales. This is why the wave-like character of matter becomes directly apparent only upon careful microscopic investigation. Just as the large value of *c*, the speed of light, obscures much of the true nature of space and time, the smallness of *h* obscures the wave-like aspects of matter in the day-to-day world.[66]

But physics again strolled down the wrong path and turned the discussion into a philosophical one.

The discourse, started in the 1920s by Erwin Schrodinger and Max Born, goes something like this: we look at a speck of something and assume we know its location. It is a mirage. What we see as a stationary object is our experience of a number of possibilities. The object exists in multiple possible states or positions (superposition), but we personally experience it in only one of these possibilities. To take this craziness one step further, it is our act of observation that causes it to exist in only one of these positions or another. Prior to our noticing the object, it existed in some fashion in all of the possible positions at once.

In this scheme, a wave is not an oscillation of some sort but instead the yes or no of a particle being one place or another. In a system where a particle has a chance of being in one place or another, this represents its yes or no-ness. The particle has a probability attached to its position, but instead of suggesting the logical expression of probability, it is supposed that the particle exists in all places at once at an amount represented by that probability. Once a conscious being observes the particle as living in one position or another, the probability spread collapses. Our observation creates the reality of the particle having a specific position. As proof of this, proponents of this theory do a mental experiment where they open a box to find the object (or absence of the object) as predicted by the collapsing probability theory and say, "See, it worked again."[67]

Cat.

If a tree falls in the woods, and no one is there to hear it, is the Copenhagen interpretation still the biggest load of crap in all of "science"? I lost five IQ points just writing this dribble down.

So then, how can particles move as waves? Their wavy essence is hidden from us because we see the macro movement. A pitched baseball travels and spins and curves on its way to the plate but nowhere in its movements do we see a wave. Its waviness is at the hyperfolded Planck vibration level. Each Planck unit involved in a chunk of matter gets pulled up into the folds, moves through its stations, and then goes out the other side, all in an instant. A given unit never holds any value position in that specific particle but, transiently, is part of the particle expression process, much like a photonic vibration.

Taking this photon/matter model one step further gives us the true essence of a Planck unit's role in the particle. When we draw a picture of a photon passing by, we are representing an event that is one Planck length in thickness. This is not a light image that would be visible. It would take a light stream a certain thickness of Planck units to reach a level that we could detect. In this thick beam of photons, there is no exactly linear transfer of vibratory information from one Planck unit to the next. Planck Unit Zero sends much of its energy to Planck Unit One directly to its right, but a small portion goes to the Planck unit at forty-five degrees up and down, the light path, plus every other Planck unit in this three-dimensional world we live in, plus a little bit of vibration to the Planck unit at ninety degrees, plus a tiny bit even reflected backward. Every Planck unit in contact with Planck Unit Zero gets some portion of its vibratory energy.

To make this equation even more complex, every Planck unit is a Planck Unit Zero. While it is sending its vibration in all directions it is also receiving vibrations from all directions. So in the overall path of a beam of light, a single vibratory Planck unit path cannot be discerned. Every vibration has a probability of existing to one degree or another in one string path or another, but no string owns the path. The detectable light we see is a swath of vibratory events transmitted in a probabilistic fashion through a cone of Planck units. As we've noted, the more focused a light is, the more vibratory energy is transferred downstream, but every light has some degree of vibratory scatter according to the inverse square law.

Obviously there are some essential differences in the effervescent photon and a chunk of matter but the model is the same. At the Planck level, what is expressed as matter really is a blur of "vibratory" events. These vibrations are both moving through the ocean background and recruiting units momentarily into their being with a slur of probability.

A baseball traveling ninety feet experiences a lot of motion and forces, but where is the waviness? It is in the vibration recruitment and transfer of energy or information from one Planck unit to the next.

A specific Planck unit is ubiquitous and nondescript. It has no destiny to be a photon or matter or graviton. During its fourteen-billion-year existence, it may have participated in all of these manifestations or, given the vastness of our universe, maybe it has rested quietly without a single disturbance. The various manifestations of Planck units blur the action of any specific unit at any specific time. A photonic or matter event has a probability of involving a particular unit, but because of the briefness and probabilistic nature of these events, it is impossible to evaluate the degree and position of participation of any specific unit at any specific Planck time. Information cannot travel faster than the universal speed limit of *c*, which is the same amount of time of unit participation, so this information is obsolete as soon as it happens.

Your fingertip, for instance, is not made up of any particular units. Given the speed of the Earth, solar system, and galaxy as we travel together though universal space, the Planck unit in your index finger a second ago are now millions of miles away. New units from the fabric are evolving into and out of your finger at a rate beyond human imagining. The units from a second ago have since participated in millions upon millions of light transmission events, magnetic events, gravity events, and particle events.

We can witness the broad brushstrokes that result from the recruitment of lots of units over an extended period of time, but this is limited by our relatively crude ability to measure at the quantum level. We may suspect that the individual Planck unit had a role in some event, but because of the extremely short distance and brief time, we can only approximate. Furthermore, as no Planck unit is distinguishable from the next, we couldn't identify a single station in our universe even if we had the capacity to measure at that level. All of our measurements of anything being here or there are probabilities on top of relative arbitrariness.

Heisenberg's uncertainty principle stands up here at two levels. Our human experience allows us to measure events at only very coarse degrees when compared to quantum time and distance. As a photon or particle event happens, it recruits local Planck units. Each unit lives in that position in that particle on a timescale of $10^{-44}$ seconds. This is way beyond any device we own to measure time or distance, and so at best we can approximate.

The real truth of Heisenberg's principle does not reside in our human technological limitations to measure, however. Even though the universe operates at quantum extremes, the universe could not measure a specific Planck unit position in time or space because this occurs at the very limit of how information transfers. In this transfer of information, even if the next Planck unit had the stopwatch, in the time it took to get to the next unit, that event has completely changed.

Every Planck unit has its specific place in the universe. As stated in chapter 4, each unit follows its own arrow of time. Its arrow is "one way": unchangeable and unstoppable.

As we've discussed already, there is a relativistic term, space-time, for the system through which these individual units experience each other. The nature of a Planck unit's experience of another depends on the distance between the two and the speed of light.

Hermann Minkowski used the mathematics of Poincaré and Lorentz, in the form of a Lorentz transformation applied to Einstein's new theory of general relativity, which treated time in a similar manner to the spatial coordinates, to arrive at a representation of space-time. In doing so, he suggested that never again should time be considered as separate from the other coordinates and that maybe someday, the artificial notion of the separation of space and time will fade away.

It is vital to the balance and workings of this functional universe that we experience the delay of space-time. Matter would not display the quality of mass, space would therefore not bend in order to manifest gravity, and light vibration transmission would be instantaneous and would therefore lose the qualities that we harness in everyday use.

As a single Planck unit participates in a vibratory transmission such as a light wave, it exists as a photon for only Planck time ($10^{-44}$ sec). In space-time, this is absolute zero time. When moving at the speed of light, all of the space-time existence is used up as distance and therefore it is not used up as time. This is our universe's expression of instantaneous.

In contrast, a single Planck unit, stationary in space, experiencing its unique arrow of time in space-time, is subject to the full extent of time and so in space-time does not experience distance.

A single Planck unit, keeping to itself, not inquiring about any of its neighboring units, does not experience space-time. It exists independently in its unique space and following its unique path in ordinary time. Space-time is how this unit experiences other units and how other units relate to it.

These two extremes are the bookends. Everything else fits somewhere in between. A baseball traveling on Earth, even with the Earth moving, the solar system moving, the galaxy moving are all traveling through space-time at normal speed and so mostly experience time's arrow. The *Enterprise* traveling at warp 0.85 will be experiencing much more distance in space-time and therefore less time.

Space-time is the theory of relativity. It is the means by which Planck units are related in time and space. It is subject to the full limitations of the speed of information transfer among Planck units. It is this apparent nonsimultaneous journey that gives us our relative physics.

# Chapter 12

## THE STANDARD MODEL IS THE WEAK FORCE

On August 9, 1945, in Nagasaki, Japan, a little over six kilograms of plutonium dropped out of an airplane. About one kilogram became fissile. In the reaction, Pu 239 rapidly decayed to neptunium 237 plus a helium atom in what is termed an alpha decay. In this process, of the original six kilograms of plutonium, only 0.9 grams of matter was converted to energy. Most of the city, along with 38,000 people, was incinerated by 0.9 grams of matter.

We think of nuclear energy as inherent to certain radioactive isotopes, but only because these are relatively unstable enough to decay and allow us to witness and even harness that energy. All particles decay, but some have such a long half-life as to appear stable. For example, even the proton decays, but it does so with a half-life much greater than most estimates of the age of our universe.

The energy is said to be related to the breaking of the weak forces within these atoms. There is energy involved in the chemical bonding of atoms to form a molecule, but the nuclear reaction is not due to bond breaking. It is due to $E = mc^2$.

Neutrons and protons share a similar space in the nucleus of an atom. Furthermore, they are very similar in mass, different almost by the mass of an electron. Soon after the discovery of neutrons, it was commonly held that a neutron and proton were two states of the same particle, dependent on whether an electron was attached or not.

It was Werner Karl Heisenberg who pondered considerably about this relationship and supposed that it was the process of transferring this electron that contributed to strong nuclear bonding and that within the nucleus this electron transfer was a continuous process that allowed neutrons and protons to exchange their identity repeatedly.

In order to do this, Heisenberg proposed a property analogous to the electric "spin" of electrons. He called this isospin. It is a bit of an unfortunate term, as it implies a physical act of spinning, but it really is only meant as a mathematical placeholder and a degree of freedom. Isospin is not actually a spin but more the additive relationship between the subnuclear particles as they interact to make subatomic particles (quarks to protons). In the mathematical handling of these quantities, they behave in a manner reminiscent of angular motion, and hence their assigned "spin" name. This is felt to be a component of the strong force that holds particles together, both at the quark level and between protons and neutrons to make a stable nucleus. The two assigned values are called "up" and "down" but again do not relate to a specific orientation.

This theory of neutron-proton interaction did not stand up to scrutiny, but the concept of isospin was considered useful and carried on by Chen-Ning Yang. He was convinced that the isospin of a nucleus was analogous to the spin of an electron. That it should therefore obey wave function and total nuclear isospin should be a value that is always conserved, that is, "invariant."

By chance, he joined forces with Robert Mills at the Brookhaven National Laboratory on Long Island, New York, and there they jointly worked on a theory where neutrons and protons were indeed interchangeable, at least in respect to how they behaved under the influences of the strong bonding intranuclear forces. The change from one to the other was solely based on a change in the isospin (rather than exchange of electrons).

For this interpretation, they extended the group theory analysis but now had two variables, with protons and neutrons changing identity. A more complex method was required, which was called the special unitary group of transformations of two complex variables SU(2). In working the details, it became evident that three new force-carrying particles were needed or predicted and they termed these B+, B-, and B0.

While this theory fulfilled many obligations in advancing QED to nuclear particles, there was a problem with the math in terms of whether these force-carrying particles were massless or massive as previously predicted and so, in respect to the strong nuclear force, this path was abandoned.

The standard model is one attempt to describe the smaller stuff. There are particles called fermions further broken down as leptons and quarks. There are six "flavors" of quarks. These come as pairs, grouped in three generations, "up" and "down," "charm" and "strange," and "top" and "bottom." They have a wide mass variation: for example, an up quark is 0.003 GeV, while a top quark is 175 GeV. Each of these can exhibit a certain characteristic that has been assigned the term "color." They are always stuck together in a pattern that yields a color-neutral pattern. According to this theory, they cannot exist alone; they must always be partnered up, and they are "confined" to exist this way. There are six leptons, again grouped in twos with three generations: the electron and electron neutrino, muon and muon neutrino, tauon and tauon neutrino (often called tau and tau neutrino). These don't need color or confinement.

Then there are the force-carrying gauge bosons. These are the supposed mediators of the fundamental four forces. The electromagnetic is mediated by the massless photon. The strong force is mediated by eight gluons, which are thought to exist under the rules of color charge. The weak force is mediated by W and Z bosons. Gravity is mediated by the graviton, which is still presently sought-after. And then, five hundred trillion collisions later, we have a Higgs boson, which mediates mass.

Oh, and then double everything, because for every particle, there is a corresponding antiparticle. Um, except Higgs.

The standard model is the present state of the art in physics. Just the name, standard model, sounds like such a highly tested and substantially proven theory that it can be assumed to exist as fact. There is some justifiable occasion to defend the existence of the first family of constituents, namely, up and down quarks. It is certainly convincingly established that protons and neutrons exist, but it should be noted that quarks have not been isolated. Furthermore, there is a theory that they can never be isolated because of this specific characterization called confinement. It was initially presumed that the subatomic particles (protons and neutrons) were made up of these smaller things, and at least mathematically, the quark model fit well.[68]

Although it may make sense that there could be more essential units than neutrons, protons, and electrons, the standard model allows for a zoo of new particles, which intuitively goes against nature's usual pattern of simplification. Dirac comments on the concept of even protons and electrons being the most basic level, noting that he feels that two particles are too many and that there must be something more essential.[69] So even though quarks are an assumption, and there are many skeptics, I do use the term freely, as if they were known to exist, so as not to dance around this issue too awkwardly.

Decades of experiments by Jerome Friedman, Henry Kendall, and Richard E. Taylor resulted in showing detail of the internal structure of protons and bound neutrons by evaluating the resultant deep inelastic scattering of an electron beam was consistent with the quark model, for which they received the Nobel Prize in 1990.

There are some mathematics that make logical sense when describing the first generation of standard-model particles, namely photons, electrons, and quarks. This math falls into the U(1) category. The U stands for unitary matrices and is part of the mathematics of circle group geometry or Lie groups or Representation groups, as discussed in the last chapter.

These mathematical manipulations are an attempt to simplify evaluations of repetitive events such as seen with circles and waves. For simplicity, the repetitive act of running around a circle is measured in degrees. One could run partially around, stop, run farther, and stop. Mathematically this could be demonstrated by adding the angles 270 degrees at the first stop to 150 degrees at the second stop. Simple math would have us add these to come up with 420 degrees, but we know that there is no corresponding angle of 420 degrees. Once we reach 360 degrees, we are back to zero, and so the corresponding angle in our example is 60 degrees. This holds no matter how many times you go around the circle; a 1,000 degree angle is really a 280 degree angle. And so it is also with waves or any other periodic function.

In the realm of this unitary group function, various circles or waves can be compared to each other, or a single wave can be slightly moved (mathematically). This could be called a phase shift or transition. It is by means of these transitions that some of the parameters of photon, electron, and quark behavior can be quantified. Although these particular studies may give us insight in the behavior of matter, they do not account for the intensely high energies of particle formation.

At the quark level, each different species would have a unique spin charge. Every baryon would be made up of three of these quarks whose spin charges are additive. The isospin of a proton composed of two ups and one down quark is +½. The isospin of a neutron is –½. A rotational transformation of the isospin property of a proton would turn it into a neutron. This macroscopically familiar term of spinning, as we've mentioned, is used solely to make us more comfortable. Up and down quarks, for instance, are said to have ½ and –½ spins, which fit nicely into equation form, but again most likely do not represent spin in our traditional sense.

Initially, the parameters of the strong force did not succumb to the mathematics of group theory, but the math was interesting nonetheless. There was felt to be another force responsible for the beta decay of certain radioactive materials that was very different from the strong nuclear force and even weaker than the electromagnetic force. Enrico Fermi termed this the weak force, and it had a very specialized niche of involvement in beta radioactivity.

Carl Anderson and Seth Neddermeyer studied cosmic rays on Pike's Peak in 1937. They found evidence of a particle that had a charge of an electron but was about two hundred times heavier. This unexpected finding did not fit with any particular theory at the time. Because of its weight in the middle of known nuclear particles, it was termed a mesotron or meson.

As more of these middle weights were discovered, they were classified by Greek prefixes Mu, Pi, and K (later shortened to muons, pions, and kaons).

This somewhat unwieldy list of new particles seemed a bit confusing, and any attempt to classify them using the previously utilized Lie strategies of U(1) and or SU(2) was not working.

Murray Gell-Mann in particular worked on this for some time until, with the aid of Richard Block at CalTech, he decided to try a Lie group with three variables—not that he knew of three variables at the time. One solution in three variables suggested eight particles to a group (family).

In Buddhism, Śrāvakayāna is the noble goal of seeking insight and controlling one's own urges for selfishness. It is an introspective study of the nature of human behavior. It is the attempt to travel the Middle Path with a goal of arhatship, or worthiness and self-awareness. It is translated for us as the Noble Eightfold Path.

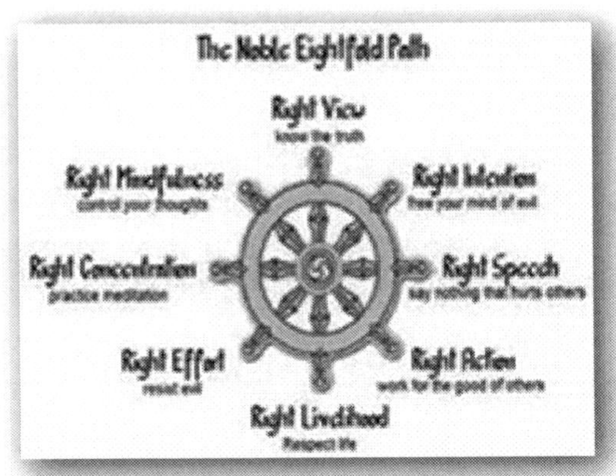


*Figure 6: The Noble Eightfold Path*

The standard model of particle physics is an attempt to enlighten us to the nature of the true underpinnings of the universe. It was called the Eightfold Way by Murray Gell-Mann. In his group representation of SU(3), his matrix pattern would predict families of eight particle varieties. The combinations were a specific ordering based on spin features called strangeness, electric charge, and isospin.

Initial attempts at reconciling total particle spin with isospin required the assumption of fractional charge values. Even the person suggesting this was uncomfortable with the idea of fractional values. With these fractional values, he was able to devise a strategy where three subunits, quarks, could combine to add up to the appropriate total spin charge. This was a major breakthrough in promoting today's theories, but at that point still had several unanswered issues. Besides the uncomfortable fractional charges, the Yang-Mills equation in use could not be completed and renormalized, which Gell-Mann was sure would be remedied soon, but more importantly, it allowed two fermion particles to exist in the same quantum state in the same locale, which is a violation of the Pauli exclusion principle, which is felt to be inviolable.

Renormalization was indeed resolved by graduate student Gerard 't Hooft using spontaneous symmetry breaking theory with the application of an additional massive particle, the Higgs mechanism.

A Lagrangian is a generic term for some parameter measurement of some feature of a system. For instance, in quantum field theory, it could be a list of measures such as spin, velocity, charge, and/or mass. The list allows comparison of several known parameters to determine an unknown or manipulation of some parameters to evaluate the change in the others. How these parameters are related is by a common factor called a coupling constant. How many dependent variables are not at a fixed value are called the degrees of freedom.

In order to offer a theory that would resolve the apparent Pauli exclusion principle, Gell-Mann, along with William Bardeen and Harald Fritzsch, decided to try an additional variable, a third degree of freedom to the equations. Each of the particles would now have a specific orientation, for which he assigned a representative color. As color was the label chosen for this quality, the theory was named quantum chromodynamics, QCD. The "chromo" in QCD is in reference to what ended up being the use of the three primary colors (red, blue, and green) to categorize them.

In moving from photons and electrons to nuclear material, QCD accounted for known and suspected particles and even predicted some unknown particles. The third parameter of QCD allowed Gell-Mann to discard the fractional charges of isospin and now assign them full integer values. The various colors could be considered various quantum states and so two up quarks could coexist as long as they were of different color. All nucleons (or Hadrons in general) were now felt to be made up of three quarks held together by gluons. Six flavors: up, down, strange (later, charm, top, bottom). Every fermion is comprised of three quarks, one of each color, creating a particle that is white (clear). Charge, spin, and mass of a hadron are determined by which combination of quarks make it up. For instance, all protons are two up quarks and one down quark, giving them a mass of $1.672 \times 10^{-27}$, charge of +1, and spin of ½.

As mentioned, I use the word "quark" as it is used in popular physics although its existence is unproven. Even if quarks do exist, it doesn't necessarily mean that all of the second and third generation of quarks listed in the standard model exist.

In particle physics, a lot of our understanding has depended on the findings derived from high-energy accelerators. Smashing massive objects into each other creates a difficult environment in which to make precise measurements, and actually both the processes of smashing and information collection cause distortions. In order to account for these distortions, there is another process that minimizes their importance by extrapolating the results out many degrees from the actual event. This is one use of a mathematical manipulation called renormalization.

The phrase asymptotic freedom was generated in the early 1970s to help explain away the unexpected findings in the local subatomic environment. An asymptote is a mathematical function of a curve as it approaches another line or, more commonly, approaches infinity. The SU(3) meaning is not so easily translated into English but can be interpreted as meaning that in the subatomic environment, the attractive forces within particles (such as gluons and quarks) increase at higher energies and as separation increases, as opposed to what should be expected in classical mechanics. In theory, the strong force of a gluon increases as it is stretched and would increase toward infinity, except that once a high-enough energy is reached, a new gluon is created.

Neither gluons nor quarks can be separated. They are confined in some particle space: hadrons. No matter how hard we pull or how much energy is applied, the bonds cannot be broken. In direct contradiction to the inverse square rule, the further these rubber-band connections are separated, the harder they pull back, up to a maximum force, at which point they break, creating a quark-antiquark meson while the nucleus snatches up another gluon, presumably out of the leftover energy of the stretching action. This turns out to be very convenient for quark theory, because if there is never an isolation of quarks, it proves confinement, and if there is, it proves quarks—win-win. There is no proof that confinement is a real thing, and so the remainder of the theory is therefore also questionable.

Three quarks go into the making of a nucleon. The expected hypothetical mass of a quark is a few MeV. A combination of three quarks would yield about five to ten MeV total. This is in comparison to the measured mass of protons and neutrons, 938 MeV and 940 MeV, respectively. Most of the mass of these subatomic particles is said to be due to the inertial resistance of the high energy bonds of the strong force, the gluons.

Particle charge was another defining characteristic. There was a famous experiment in 1909 performed by Robert A. Millikan and Harvey Fletcher where the velocity of charged oil droplets was measured for various-sized and -charged droplets. This was done in a chamber where a variable electrical field could be applied. By applying this field, the oil-droplet velocity was slowed. By adjusting the field strength, the oil-droplet velocity could be slowed to zero, the droplet perfectly suspended in the chamber. With an understanding of the gravitation force versus the applied electrical force, the electrical charge of the droplet could be determined. This value was always a multiple of some number, and so that common number could be determined. This constant was found to be approximately $1.6 \times 10^{-19}$ C. This fundamental charge was felt to represent the charge of a single electron. This also helped support the burgeoning theory of subatomic particles and that charge was quantized.

Mesons are intermediate-weight particles initially detected in cosmic rays. As mentioned earlier, they have been classified in the standard model as muons, kaons, and pions. These have since been demonstrated in particle accelerators. The Mu mesons and the K mesons in particular were of great interest in developing the theory of weak interactions.

At least the fact that they occur naturally and were detected from cosmic rays adds something to the validity of these theories, as opposed to measuring hadronic debris from an accelerator and attributing natural attributes to artificial fragments. These particles have a half-life on the order of $10^{-8}$ seconds, allowing the breakdown remnants to be further investigated.

In these decomposing events, some starting particles are known, and some resultant particles are detected, such as protons, neutrons, electrons, and mesons. However, to account for some of the behavior of these events, such as change in charge, spin, or inertia, other particles were assumed to make up the difference.

One set of these were ghostly particles, called neutrinos, which are massless or nearly so and are also of neutral charge, making them essentially undetectable. There are some techniques to detect these things, but for some it is still not clear if a neutrino is an actual particle or some reverberation of an event.

Another set of particles, which exist in theory and fill out positions in the standard model, are the force-carrying bosons of the weak force. Although these are felt to have mass, and some are charged, they exist so briefly that they cannot be directly visualized and verified.

W and Z bosons are felt to be the elementary standard particles mediating the weak force. In theory, these should be massless, but it is thought that through their interaction with the Higgs mechanism, they gain mass. As a rationalization for this discrepancy, it is assumed that these particles need mass to exert their influence at short range, within the nucleus. They are presumed to be relatively large, one hundred times bigger than a proton. They have half-lives of about $3 \times 10^{-25}$ seconds, and so again they have not been directly visualized. It is theorized that these particles specifically mediate the nuclear fission of particles with emission of electrons or positrons. They are said to mediate neutrino absorption and emission. What the heck do Z and W bosons do, exactly, in a mediation?

The 1957 Nobel Prize was awarded to Tsung-dao Lee and Chen-Ning Yang for their statistical prediction and description of weak interactions. The particles mediating the weak force are best known for generating Nobel Prizes. For instance, Carlo Rubio and Simon van der Meer were awarded a Nobel Prize $10^{-25}$ seconds after isolating a signal interpreted as a Z boson. The Z, named by Steven Weinberg, is variously described as representing the last needed particle to complete the standard model versus Z for zero charge. Z0 is said to participate in weak force interaction by causing the scattering of electron neutrinos off of up and down quarks by exchange of this electrically neutral quantum, resulting in no change in neutron or proton flavor.

There is no collection of W bosons in some laboratory somewhere to investigate. The half-life is so short that it doesn't move beyond the radius of the smashed proton that created it before it disintegrates. It has never been specifically identified. This particle is presumed due to a lack of momentum in one direction that needed to be accounted for. At the very least, there is minimal scientific proof of W's existence.

A W is a creation of a computer modeling program. No one has seen an actual W. What has been interpreted by the computer as a W is a collection of circumstantial evidence. First, in this great background of background noise, a one in one-hundred-billion event was captured. This event was not a single W particle; this W theoretically lasted only $10^{-25}$ seconds, way too briefly to come close to a detector. Instead, what was measured was a muon, a particle comparable in charge to an electron, but heavier. It was the behavior of the muon that was felt to account for the new particle. Specifically, the inertia could not be fully accounted for. The missing inertia was in an amount consistent with the loss of a neutrino. Again, the neutrino itself was not measured, as it is essentially not measurable with its low or no mass and neutral charge. Its existence was implied by the subtle change in momentum of the muon. This is the hard evidence of the existence of a W boson.

The weak force is not necessary. All mass carries radioactive potential energy in the amount $mc^2$. All particles are held together entirely by a single nuclear force with an energy proportional to its mass, which could be termed the strong force. A large atom, uranium for example, with more protons and neutrons, is more unwieldy. The more unwieldy it is, the more likely it is to fall apart. This is not because a different glue is used to hold the unstable parts together. The pulling apart of strong nuclear forces is plenty enough to account for the strange goings-on in radioactive decay.

It is like the puzzle game Jenga. In this game players take turns placing blocks of various size and shape on top of each other, building a narrow skyscraper. As the first few rows of pieces get laid down, they are quite stable, following the rule of gravity to keep them in place. As you place pieces on row twenty, the whole thing wants to topple over. This is not because strong gravity works on small puzzles and weak gravity works on unwieldy puzzles.

If you were to take a large, thick steel plate (like you might see covering some roadwork) and try to lift it, you might find it impossible because of its immense weight. Now balance that plate up on its edge, standing straight up. It would take only a gentle nudge (a weak force) to topple it over. The same physical properties are in effect, but the steel plates are in two very different stability states. Isn't it odd that the most radioactive elements all have high atomic weights and that all high-atomic-weight elements are the most radioactive? Some of these elements last only briefly under laboratory conditions. It is amazing that Zs and Ws know to pick them out specifically for decay.

It is equally amazing that these magical weak-force mediators, acting thousands of times in each gram of uranium 238 per second, find their target atoms at such a precise rate, and yet they are not abundant in nature. They know to act at a different rate for radium. Again, these bosons are not found in nature. They are only discoverable under the extreme conditions of particle accelerators. If they are responsible for innumerable reactions per second, we should be able to place our detectors out in the natural environment and see a massive bombardment of W bosons. We should be able to isolate uranium from natural Z and W bombardment by encasing it in lead deep in the tunnels at CERN and keep it from ever decaying.

These fancy particles, which exist only in particle accelerators to perform amazingly complex functions, do not meet the minimal sniff test for reality. The particles created in accelerators with exceedingly short half-lives therefore do not genuinely exist in nature. Because accelerators smash particles into fragments, there is an implication that nucleons must be made of smaller stuff. It does not necessarily mean that these specifically smashed bits are that stuff. Even so, there must be smaller stuff.

The recently discovered particle was also too short-lived to actively participate in any event. It is not a mediator of an all-pervasive field. Its only real purpose is to make it so that the Yang-Mills equations don't have infinities.

By having $10 billion of backing, the high-energy physicists have, by default, been given the authority to decide on what is acceptable physics science and what is to be disregarded. This is somewhat similar to the role of the church during the inquisition.

These particles are talked about with complete confidence, but they are only theoretical, and many of them make no sense. They are said to have been discovered, but many have half-lives way too short to ever be observed. The non-physics world would be surprised by the scant evidence by which these particles are "proven." It would be good for us to just stop making this stuff up.

**Conclusion:** The word "particle" is used as if we had a clear notion of what is meant, but some of what we call particles may be no more than debris, or the flash on a detector screen, or sometimes an unexpected void in the inertia of the debris. In this sense, a particle is a collection of properties—sometimes implied—rather than something we can hold in our hands. None of these is even close to being proven, and all are only theoretical. Planck units are proven by every experiment ever done and every moment in every person's life because everything we do is through them.

# *Chapter 13*

## HOW WE BECAME US

The universe is flat. This statement has great metaphorical value, as it sounds so similar to the old fallacy that "the Earth is flat," but it is a bit misleading and reflects on how little we really know. The shape of the universe is not known (it is approximately spherical, by the way). What is really meant by the flatness statement is that, in our universe, Euclidean geometry seems to behave. Our universe has linearity in its infrastructure. Our quantum fields appear to transmit information in a nonbent manner. In our space, for instance, the three angles of all triangles will always add up to 180 degrees.

We need to remember, however, that even in a bent space where the angles should not add up to 180 degrees, if the protractor is bent by that same amount, then we could still have the appearance of 180 degrees.

The flat universe is actually a mathematical statement regarding the amount of mass, energy, and gravity of the universe. It is a statement of how different models of the universe would behave under different quantities of this stuff.

If the universe is expanding, we picture expansion occurring somewhere way out in space. Maybe we picture expansion of the leading edge of the universe and maybe some of the galaxies out near its perimeter. Maybe we picture the space in between galaxies, or maybe we could even assume some expansion within our galaxy between the individual stars.

The universe, being a noncognitive entity, would not know to behave one way at its perimeter and another in our local experience. If the universe is expanding, it is doing so at every point. There would be a continuous expansion process between all points. The space between all atoms in every molecule in your body would be expanding at a rate in proportion to the overall expanding universe. Each atom itself would expand, as would its component parts, at exactly that same ratio. Fortunately, every yardstick is also expanding at precisely the same rate, so, relatively speaking, there is no perceived change. Our point of reference stays comfortably stable.

If tomorrow I am a millimeter taller, but every other object around me increases proportionally, I will not be able to perceive the change. Even if I am a meter taller or a kilometer taller, if everything else changes to that same degree, I will not sense it.

If every distance doubled overnight, then light would have to travel twice as far by yesterday's standards. But if every

Planck unit increased proportionally, the perceived speed of light would stay the same. It would still transmit from each smallest possible space to next space in exactly Planck time. Light from the Sun would still reach us in about eight minutes.

In an expanding universe, it appears that all galaxies are moving away from each other. The rate of this movement can be estimated, as can the expansion rate of the universe as a whole based on Hubble's constant (in theory). Based on these observations, it appears that the rate of expansion is accelerating. All galaxies may be moving at their fixed rate, as is dictated by conservation of inertia, but the surrounding Planck field within which they are embedded also is expanding at that same rate. What would appear to an observer as a "fixed" point when observed in close proximity would, from a more distant vantage point, appear to be accelerating.

Picture two cars are driving side by side at sixty miles per hour. One car accelerates by one mile per hour per minute. In ten minutes the second car is moving at seventy miles per hour and pulling away from the other car at an additional ten miles per hour at that instant. Both drivers would agree to these observations.

In another experiment, two cars are driving at a constant rate of sixty miles per hour but now are separated by a mile. They are now driving on a magical road that is continuously stretching. It stretches at a rate of 1 percent per minute. Each driver looks at his speedometer and confirms that he is going sixty miles per hour. Each one looks out the window and confirms that the trees appear to be passing by at the same rate.

Each of them also notices that the distance between the two keeps increasing, which is consistent with acceleration.

The Earth is embedded in an elastic sheet called the universe. As the universe expands, and the sheet stretches, it does so in a diffuse manner. As a model, you take an elastic sheet and draw two dots at 1 cm and 5 cm. Over one second, stretch the sheet so that now it is doubled in length. Dot number one has moved to the 2 cm position and so has traveled 1 cm per second. Dot two is now at the 10 cm position. Having moved from 5 to 10 in a second gives a velocity of 5 cm per second. Both dots were part of the same elastic stretch, which is a uniform process, but dot two appears to be going faster. The red shift of this process would suggest an acceleration.

Wherever we look, we are seeing red. So says Hubble. With few exceptions, all of the galaxies except those in our own local clusters seem to be receding from us. The galaxies farthest from us seem to be receding faster than those closer.

On the other hand, maybe the universe is not changing in overall size. What if the universe existed at a fixed size, pinned by a static structural scaffolding, while its contents are shrinking at a rate consistent with Hubble's constant? As each Planck shrinks, it creates the impression that it is farther away from the edge of the static scaffold. Only kidding. It helps if you lived through the 1970s for that to make sense.

In the usual model for the expanding universe, acceleration is not thought to be a steady state but instead had a burst six billion years ago. This may have been an actual acceleration or an apparent acceleration. If we are, for instance, sitting at the 8.6 billion light-year position from the Big Blast epicenter, in a 13.7-billion-year old system, as we look back toward the blast, we see apparent steady expansion and look out and see accelerated expansion. We could be miscalculating, because somewhere in our primitive souls, we need to believe that we still maintain our position in the center of the universe. The beginning-of-time calculations don't start on Earth.

In our expanding universe, it appears that this process is accelerating for that same reason. If a particle, or a galaxy, for that matter, is moving away from us, but is doing so as every Planck unit is enlarging, it will appear that the particle is accelerating. The very expansion amplifies the appearance of its own expansion. If you were standing on that other galaxy, it would appear that the telephone posts are going by at a steady rate.

There is a general consensus that the radius of the observable universe is 46.6 billion light-years due to the accelerated expansion model. Although we view the singularity as a point, the stuff on the outer surface of that point (I realize that makes no sense) would suffer the full power of the sudden expansion and now be at the outer limit of our universe. The stuff at the center would be pushed by vectors in all directions, leaving it relatively stationary. If, as some believe, there was a specific event in a specific point and time, there should be some central stationary point. We should be able to locate what exists there now. It would make sense to find something existing one light-year away from the original point. The cosmological principle, Edward Arthur Milne says, is that all galaxies move about the same speed, but the relative speed is proportional to the distance between the stars. Proven by Hubble, this theory can be stated in another way: if we know or suspect that every galactic object is moving at approximately the same speed in the great framework, then the apparent speed based on red shift will correspond to the distance from the observer in the ratio of apparent speed/real speed. If true, then is the stuff at point zero especially slow-moving, or it has been elsewhere and is now just passing by again?

In the early years after the Big Bang, the elements that would later become mass would have occupied a large percentage of the small universe. At the end of one year, was the radius of the universe one light-year, or was it more a process where the immediate contents of the Big Bang took time to fill the enormous, awaiting scaffolding of the present universe?

If the Big Bang's singularity was the entire limit of our universe, then all of its entirety and the scaffolding was contained within it. If this entirety is made of Planck units, then these units necessarily would have increased in size along with the universe. And yet, despite this continuing change in size, $c$ and $h$ don't change. The implication that the Planck units were proportionally smaller would suggest the universe has always been and will always be measurable at 13.7 billion light-years in radius. If someone from today's universe could magically measure the one-year universe by today's standard, we would calculate two light-years in diameter. From within that much smaller space, we would measure 29.2 billion light-years.

So the universe is flat. The first time I was taught this many years ago, it was taught to me as we commonly understand flat. It is a very misleading term and is probably often incorrectly taught and will be passed on incorrectly to future generations. The term suggests that maybe our universe resembles a sheet of paper. This incorrect model is then further solidified at the end of the lesson by showing how it predicts Euclidean geometry. The teacher at this point draws a triangle, showing that due to flatness, the angles add up to 180 degrees, as they should. We are taught this, tested on it in school (it is answer B), hold it deep in our souls, and force our students to carry it on to the next generation. Ask a physicist how the universe is flat, and the answer invariably is Euclidean geometry...blah, blah, blah...180 degrees.

As mentioned, the flatness of the universe is actually a mathematical model estimating the rate of expansion of the universe. It is the intermediate solution between an open universe with expansion that accelerates into oblivion and a closed universe that collapses back on itself. The flatness is an unfortunate phrase coming from the geometric model of these equations. A closed universe resembles a sphere. An open universe resembles a saddle. In either of these situations, a triangle sitting on this irregular surface would end up having its three angles add up to more or less than 180 degrees. The math of the intermediate solution is consistent with Euclidean geometry: flat.

If the universe were spherical or saddle-shaped, then my protractor, which was created in this curved universe, would be distorted to the same degree. Even if created in a "flat" universe and transported here, it would become distorted by the curved universe. We would still measure 180 degrees, whereas someone in the nearby flat universe with his or her supercool telescope would see the distortion. Even if we could measure the distortion ourselves, in a universe so immense, the distortion could be in the fiftieth place (i.e., 179.999…9°) and therefore hardly measurable with our meek equipment.

Within a year of publishing his general relativity theory, Einstein also published his work attempting to define the total mass and gravity of the universe. This result was a bit resistant to general relativity as his calculations yielded a closed universe. The attractive forces of gravity would not allow the large, static universe that was the present standing theory. Given the amount of mass projected, gravity would cause all of the particulate mass of the universe to clump together. The pressure was so great, it would collapse the universe into an infinitely small singularity. As we've already discussed, this was the origin of the cosmologic constant—a constant force designed to overcome the contracting pressure of our universe's total gravity.

The idea that we fit into a flat pattern of the universe expansion gives the calculations of the Big Bang its power. A certain amount of energy was present at the instant of the Big Bang. Fortunately for us, it was the correct amount of energy—the critical energy. Any more, and the universe's gravity would have caused a collapse back to nothingness. Any less, with not enough mass/gravity to allow stable expansion, and everything would have blown off to infinity (at least mathematically). The math used to define the "too much" scenario is that used to define a sphere. Again, this is termed a closed universe, as it would collapse in on itself. The math that describes the "too little" mass scenario is hyperbolic saddle-shaped. It is termed an open universe. In the middle is the baby bear. Just the right amount of energy/heat/mass/gravity. It allows expansion without a catastrophic obliteration.

According to Weinberg, at 0.01 seconds post-Bang, our new universe existed at about one hundred billion degrees and about four billion times the density of water. Assuming the word "temperature" is even the right expression for this level of kinetic energy, at this temperature, our universe would have consisted of electrons, positrons, photons, and neutrinos. All of these would have started at approximately the same amount in this high-energy plasma. In addition, there would have been lesser amounts of protons and neutrons in the ratio of one to one billion electrons.

The universe quickly cooled and was

* at 0.1 seconds, thirty billion degrees;
* at 1 second, ten billion degrees;
* at 14 seconds, three billion degrees; and
* at 3 minutes, one billion degrees.

It is Weinberg's theory that at this temperature, protons and neutrons could gather together to form atomic nuclei. It did this in a ratio of 73 percent hydrogen nuclei and 23 percent helium nuclei. In addition, by this time most of the electron-positron annihilation would have been completed, leaving the excess electron count at approximately that of the protons. Even though these coexisted in the universal stew, they would not combine to form atoms for another several hundred thousand years.

As mentioned, matter formation (nucleogenesis) requires energy at a dose at least as great as the resting mass of the particle, according to the equation $E = mc^2$. Therefore, heavier elements would necessarily require higher energy and could have been formed in the very earliest moments of the universe, but because of the intense radiation at that point, they would have been blasted apart as soon as they were created. By the time the radiation was low enough to allow stable heavy-element formation, the temperature was no longer high enough to create them. The result was the aforementioned ratio of 73 percent hydrogen and 23 percent helium, with lesser amounts of deuterium and lithium. This is all there was and accounts for all there will ever be.

Rather than spreading out diffusely through the universe, the gravitational effects of the early particles encouraged clumping into great clouds called pillars. These pillars were so large as to dwarf even the largest galaxy. Within these pillars, there were areas of extra "clumpiness" where the mass and gravity and pressure would cause intense heat. An equilibrium would occur between the temperature and collective mass of hydrogen, and this clump is the birth of a star.

Some collections are too great to counterbalance the temperature. Here the gravity outstrips the opposing kinetic energy, and the structure collapses on itself. This is the birth of a black hole.

A black hole of a large enough size has an intense attractive gravitational force to pull the nearby collection of stars together. It does so in the repeating pattern we see in nature, as the form of a spiraling disc of orbiting structures—in this case, stars. This is the birth of a galaxy.

At this point we still only have hydrogen and helium in our universe. The stars fuse hydrogen into helium and give off radiant energy, which adds to the temperature. This nuclear fuel does have a limit, and as it starts to run out, the core of that star begins to collapse. In a star of great enough size, the collapse will result in a higher temperature, at which point the helium takes over as the primary fusion fuel.

Helium fuses to form beryllium, carbon, oxygen, sodium, and magnesium. As the helium is used up—and, again, assuming there is enough mass to sustain the next higher temperature level—the star contracts again until it reaches a temperature where oxygen can fuse into silicone and sulfur. With some of the leftover helium, argon, calcium, fusing to nickel.

Up to this point, fusion of nuclear material has released the energy to sustain this process in the solar furnace. Adding any more neutrons or protons beyond this point actually takes additional energy and thus is not favored. Some of the elements that were skipped in this process now begin to emerge as the less stable nuclei start to decay in natural fission. Fission does release energy and is now favored as the reaction to sustain the internal temperature to keep the gravity of the star collapsing on itself. Much of the unstable nickel-56, for instance, decays to the much more stable iron-56, and so we have an abundance of iron.

A point is reached when there is no possibility of thermal equilibrium, and the star collapses completely and violently on itself. This collapse is accompanied by a great increase in density and a resultant great increase in temperature. The pressure and temperature are so intense that now all the other heavier elements form. This process is not sustainable and ends quickly in a great explosion we call a supernova.

A supernova showers the universe with its particle collection and is the source of everything that we know and love.

In 1930, Wolfgang Pauli proposed the existence of a small neutral particle emitted in beta decay in order to conserve energy, momentum, and spin. As we've noted, a few years later, Enrico Fermi coined the term neutrino, which, loosely translated from Italian, means "little neutral one."

Most of the neutrinos striking the Earth come from the nuclear decay of helium to deuterium from the Sun's furnace. An estimated sixty-five billion neutrinos strike every square centimeter of the Earth's surface every second.[70] This bombardment goes totally unnoticed, however, because the neutrinos have minimal interaction with ordinary matter. In fact, almost every neutrino passes through the entire thickness of the Earth without the slightest impedance.

Our instinct would tell us that this particle must be very small to pass through all that thick matter without hitting something, but that is actually not the issue. No matter how small a neutrino may be, one would think that it would encounter atoms immediately. The idea of passing through thousands of miles of densely packed stuff to emerge on the other side completely unscathed is unfathomable.

The two partners in this event are the neutrino, which has no charge, and the physical structure of all of the atoms that make up the Earth. Both have particular characteristics that allow the passage through such a vast amount of matter. The neutrino has no charge and so does not experience attractive or repulsive forces of charged particles, but it is the overall structure of atoms that has the biggest role in allowing this particle passage.

We all have the image of an atom distributed by our junior-high-school teachers on Ditto paper. Here the nucleus took a significant portion of the image, and electrons were little accessories buzzing about. In actuality, the nucleus is best represented by a barely visible speck while the field of influence of the electron "cloud" should take up the remainder of the page, or better still, the remainder of the classroom.

While it is essential that the neutrino is not interactive, the truly amazing fact is that each atom is made up almost entirely of empty space. The stuff of an atom occupies only a tiny fraction of its volume. This is so much the case that even given the almost inconceivably large number of atoms that a neutrino would have to pass in its travel through the full thickness of the Earth, it is unlikely to ever strike the "speck" and will freely pass through the "space."

The speck and space model is vital to the structure of the universe. The large electron field makes up the entire exposed surface of an atom. It shields the positive core from exerting its influence on the surrounding environment. While the total charge of an atom is neutral, the outside world only experiences its negativity. It is the electron field and its stability patterns that allow for molecular bonding, but also it is the electron field that distinguishes one atom's space from another atom's. Due to the repulsive force of one cloud of negativity repelling another, atoms have a size and structure much greater than the component parts.

This size and structure are vital to the makeup of the universe. If an atom were not so fluffy, the component parts could occupy a much smaller space; neighboring atoms could be bunched up more closely. The density of matter would create such gravitation force as to crush everything into black hole-ness.

On the other hand, if atoms are mostly space, then shouldn't one atom should be able to pass right through another with a low likelihood of interference? When you poke someone in the ribs, your finger should go right through. When you try to stand on the surface of our planet, you should sink right through to the core. Try to put on clothing; they should slide right through you to the ground, and beyond.

We are made almost entirely of empty space. The nucleus of any atom occupies a minuscule percentage of its particular field. It is separated in distance to its neighbor by, at the very least, the radius of both atoms' fields added together. If my finger is mostly empty space, and my abdomen is mostly empty space, then the nuclei of my finger have a low likelihood of hitting a nucleus of my abdomen, and so one should easily pass through the other. Obviously this is not the case. It is the lowly, wispy electron field that accounts for our substance. The additive negative charges of all of the electron fields (the portion of the atom that is its outer exposure) repel the negative charges of all of the atoms that make up my skin and internal organs. Furthermore, it is that same negative charge that repels us from the surface of the Earth rather than allowing our minimal-volume nuclei to keep us from gliding freely to the Earth's core. It is not the meat of us that makes us structurally sound; it is the negative electron field that offers appropriate size and structure. It is our intrinsic repulsiveness that makes us who we are.

# *Chapter 14*

## THE ABYSS

Each of us has a sense of our self. We believe in our existence. We are self-aware to some degree or another. We have plans; we have memories. Our ambitions help us get closer to our goals. Our goals are vitally important to us. Our bad habits drive us crazy. We see some injustice on TV about someone whom we have never met and who lives thousands of miles away, and we are, for some unknown reason, outraged.

There are some pretty fanciful and even extreme ideas on the nature of our existence. It is most likely that we exist and do so in a fashion that we share as our common experience. Furthermore, our existence may be somehow important to the universe. Though we may not be the geographical center of the universe, it seems to some of us that we are the reason for it. Despite billions of years since the Bang, our present time seems to be the most important time.

On the other hand, the Earth is really a mere speck of dust in the vast cosmos of the entire universe, and not only our present time on Earth, but also even the collective time of all human existence on Earth, is a blink of the eye in the story of the universe. By this estimate, it is hard to express in words just how truly insignificant we are.

In our sense of the universe and all that it contains, in our common three-dimensional experience, it is hard to fathom the comparison of the great expanse of the cosmos versus the lowly quantum unit that makes everything in it in the same thought, but it is one of my favorite exercises.

Man is somewhere in the middle. Compared to the size of the entire known universe, Earth and man are approximately the same size. Compared to a Planck unit, Earth and man are approximately the same size.

As I have mentioned, a Planck unit is about $10^{-35}$ meters in diameter, which is a totally meaningless number to us. The probable diameter of the wispy electron is on the order of $10^{-17}$ meters. A proton is about $10^{-15}$ meters. The diameter of a hydrogen atom may not actually be definable, as electron field behavior makes it difficult, but it is probably on the scale of $10^{-12}$ meters. Those tiny numbers are nearly meaningless to us, but the differences so far are immense. For scale, if a hydrogen atom were the diameter of a football field, the nucleus would be about one millimeter across. If a hydrogen atom were the size of the universe, a Planck unit would be about a millimeter.

Viruses vary in size, but the smaller end of the scale is about $10^{-9}$ meters. A human white blood cell is ten times larger. We have a wide variety of cell types doing all sorts of functions and having a wide variety of sizes. A motor nerve cell to our foot will be over a meter in length. We are made up of 1,000,000,000,000 cells, give or take. We are on the order of $2 \times 10^0$ meters tall. There are about $7 \times 10^9$ of us now and probably about $7 \times 10^{11}$ of us ever.

Mount Everest is $8 \times 10^3$ meters tall. The diameter of the Earth at the equator is about $1.2 \times 10^7$ meters. Light could travel seven times the circumference of our planet in one second. If we think of a town three hours away, that's about two hundred miles of highway driving. We wouldn't drive that far typically without significant forethought and planning. It is a pretty significant distance in our everyday lives. Now imagine driving that same distance straight down into the Earth: two hundred miles. So think of things as large as the Earth, and yet in cosmic terms, this is an insignificant speck.

The moon is considered approximately the outer limit of Earth's gravitational significance and is about $4 \times 10^8$ meters away from us. It takes light 1.3 seconds to go that far.

The Sun is about $1.5 \times 10^{11}$ meters from us, or eight light minutes. Pluto, our ninth planet, is $7.5 \times 10^{12}$ meters from us. Our solar system is about four light-years in diameter, which is also about the distance from us to the next nearest star, Proxima Centauri.

The Milky Way is about one hundred light-years across and houses about three hundred billion stars. One galactic year takes about 250 million years, so Earth has made the full trip a little over eighteen times. Our galaxy is a member of a local group of a couple dozen galaxies, all under the influence of a mutual gravitational force, including the Andromeda galaxy.

The Virgo Supercluster contains our local group cluster of galaxies along with a hundred other galaxy cluster groups. It is maybe 110 million light-years across. The Pisces-Cetus Supercluster Complex holds our Virgo Supercluster, along with possibly sixty other superclusters or possibly tens of thousands of galaxies. It is one of the largest structures identified thus far in our known universe and is about 1.4 light-years across. It is about a twentieth of the diameter of our known universe.

The extent of our universe is not really known. We have traces that lead us to think that it is twenty to forty billion light-years in diameter. From a philosophical standpoint, it only makes sense as a finite quantity. It is suspected that there are about ten billion galaxy superclusters in our universe, hundreds of billions of galaxies, 30,000,000,000,000,000,000,000 individual stars.

There is a school of thought that describes the universe as an immense relay of information. What we sense as reality is our interpretation of this information grid. It is our interpretation of very simple data, the on and off, the ones and zeroes, the bits that are coded and transmitted at the smallest scale. In this theory, the computerlike quality of this information is all there is. That is to say, we personally would be no more than these information bits that define the rest of the universe. It is not clear in this picture how there is room for human consciousness.

Modern computer games take an enormous amount of memory space to create the lifelike images doing battle. These images are complex and give an appearance of reality. If one were to look at the computer code, it would at its most basic level be zeroes and ones in a specific pattern, a massive list of simple binary digits. This code would not hold any outward value in relaying these images to our brain.

If this theory of simple binary digits is all there is, where do the bits reside? Well, of course, as everything is Planck units, they would have to reside there. The information signal is transmitted as bits, unit by unit longitudinally in what could be thought of as an information string. A collection of these information strings recruiting the Planck units directly lateral to this index string traveling as a wave package gives the overall message structure. These, for instance, could represent the compaction and rarefaction of a transmitted light wave or a complex image that we interpret as matter. In this sense, particles do not have an ultimate essence, they seem to maintain a constant form and presence, but they are part of a process that is the continuous flux of the universe.

Pretty much all of us know that a hologram is a three-dimensional projection. Another way to think about it is that it is a complex construction of information inside a projection system and also the resultant projected image. The complex list of zeroes and ones and the image are two separate forms of information completely defining the same thing, but completely different in their outward appearance. If we were to look at the pages of digital code responsible for the image, it would not carry much meaning. It is not until we view the result that this information takes its shape. Even so, this would suggest that the information list is real, and our image is imaginary.

The idea that the universe is merely an interpretation of binary information is likened to a hologram. We are quite used to our experience of the hard objects that fill our world. It is a big leap to consider all of these objects as incidental manifestations of a more primary information system. When we look at what we presently call a hologram, we would be looking at the universe information list that creates the image of Carrie Fisher that was then encoded into another information list to create an image of Princess Leia.

Furthermore, while we learn about entropy as the disorder of a system, it may be that the essence of entropy is an account balance of all of this information. If that were the case, entropy would no longer be a result of the system but the full descriptive account of the system.

The holographic theory is an offshoot of string theory in the study of black-hole behavior. As I've stated already, the study of black holes is really the study of the theoretical event horizon rather than the nugget of mass at its core. For instance, we measure the energy and entropy of stuff as it falls into a black hole, and we interpret it as if the information were written on the surface of the event horizon. This can be thought of as the interpretation of elemental information on a two-dimensional surface. For instance, in looking at the thermodynamics of this system, we can estimate the entropy as two-dimensional surface fluctuations.

As an extension of this theory, it has been conjectured that the universe behaves in a similar manner. The maximum entropy of this system is proportional to the square (surface area) rather than the expected cube (volume) of the system. In an attempt to explain a three-dimensional world with equations having nine dimensions, one could treat the volume of the universe as a black-hole equivalent with an event horizon. The imagery inside this volume is encoded on the two-dimensional boundary of the volume, the event horizon.

Rather than reality as we all expect it to exist, we see images that are projected onto the world screen but in essence are the complex configurations of these information bits in the background. In *Wholeness and the Implicate Order*, David Bohm reminds us, "Quantum Theory contains an implicit attachment to a certain very abstract kind of analysis which does not harmonize with the sort of indivisible wholeness implied by the theory of relativity."[71] The universe can, in fact, be defined mathematically at the Planck level by these bits, and it agrees very well with quantum physics as well as relativity.

The information storage and transfer unit is the Planck unit. All the mass, energy, and momentum of the universe is encoded into the motion picture of Planck units. The total of everything that is still only recruits a very small percentage of total available units. Entropy is the increasing distribution of information on more and more units. Complete entropy would be the even distribution of information on every single unit. It would be featureless. It takes the asymmetric distribution of information to impart structure.

We all acknowledge that a hologram is imagery and therefore less than real. If our experience is holographic, the life that you have worked so hard to arrange—your past, present, and future, as well as the future of your kids and their kids—is the hologram. Every minute detail of every single action, and even every thought, would be a projection of collections of essential information bits, the imaginary and complex projection of *Grand Theft Auto* played on the cosmic scale.

If the entire universe is merely a projection of information in binary code onto a screen that we call reality, where is our place in this? What is human consciousness and determination? Are we real? Do we exist? Are we the collective computer dream of one conscious entity? Are there really seven billion people in the world, or just one, you, the dreamer (god?)? Have you constructed this very elaborate dream called the universe to fill your otherwise void eternity? Be careful that you don't awaken with this self-realization, or this whole dream might vanish, at least until your next fifteen-billion-year nap.

There is an idea called the principle of mediocrity that suggests that it is wrong for humankind to think that we have some central role and purpose. This is in response to an extension of the ancient geocentric belief that Mother Earth is the center of the universe. It is still believed to this day, and bolstered up by some cosmic measurements, that it seems that the universe extends out to a similar degree in every direction from us. This would imply that Earth enjoys a fairly central position in the great framework and is either an amazing coincidence, given the vastness of the universe, or divine intent.

The likelihood that we are not in any central position is a theme derived from the days when it became clear that we revolve around the Sun, from which we get the descriptive phrase "Copernican modesty." What this may suggest is that most likely, the universe extends beyond our fourteen-billion-light-year estimate, but the flashbulb of some great cataclysmic event has left its fourteen-billion-year timestamp, and this limits the Polaroid of what we can see.

This book is meant to be more a philosophy book rather than a physics book. None of this is fact, and the enormous number of assumptions that it is built upon makes it impossible to claim any of it as truth. It would be very arrogant to think that any of this Planck unit theory is correct, but there may be something to be learned from the exercise. Even if it is 95 percent wrong, it would still be such a vast improvement over the standard model, which is 99.9 percent wrong. Unfortunately, I just don't know which 5 percent is correct, but maybe at least this is a new and much more valid starting point.

This general theory is based on the experimental results of Einstein, Lorentz, and Planck. In Max Planck's Nobel address, he remarked:

> Either the quantum of action was only a fictitious magnitude, and, therefore, the entire deduction from the radiation law was illusory and a mere juggling with formulae, or there is at the bottom of this method of deriving the radiation law some true physical concept. If the latter were the case, the quantum would have to

> play a fundamental role in physics, heralding the advent of a new state of things, destined, perhaps, to transform completely our physical concepts which since the introduction of the infinitesimal calculus by Leibniz and Newton have been founded upon the assumption of the continuity of all causal chains of events.

It is imperative that we go back a century to get ahead of the evils and inconsistencies of present-day high-energy physics. For one hundred years at least, we have been heading down the wrong path. It is now a steep downward mountain path where no one desires to turn around. Everything wrong about this path is either ignored or manipulated away, but very few are willing to admit that the path was a wrong choice from the beginning. In fact, the few voices that are heard are considered heretics by the high priests of particle physics and sentenced to excommunication and purification by death to the physics world.

This mountain path is especially slippery now as we have so much invested in the LHC and in the thousands of grants and the tens of thousands of particle physicists who have devoted decades of their careers to this one hope. For them, turning back is impossible. And so it is for all of us, because the powers that be set the agenda and tell us the scope of allowable truths.

First, we are not going to find some "theory of everything" by regurgitating the teachings that haven't worked to date. The fact that it has escaped our grasp thus far should be evidence enough that we are missing a fundamental clue.

Second, devoting your life's work to tweaking an already existing five-age equation is too focused and misses the big-picture-look necessary for an overhaul.

Many, including David Bohm, tell the story of Ptolemaic study of the circular patterns of planetary motion around the central Earth. This was the dogma of the time, and so, as more data emerged and contradicted the dogma, adjustments were made to include at first elliptical orbits; and as these became unworkable, little miniorbits called epicycles were invented to maintain the theory—a fudge factor so that Mars, for instance, was revolving separately around the Earth. This unfortunately (but obviously) did not fulfill mathematical findings, and so a new theory had these epicycles revolve around an arbitrary point in space called an equant. All it takes to make a wrong theory workable is a little more math.

As these renormalizations became more complex, it should have become clear that the theory was wrong. We employ these same complex renormalizations in today's standard theory of particles. This is a clear statement that a paradigm shift is needed—or, as Bohm calls it, a New Order.

I agree with the words and sentiment of Alexander Unzicker in his book *The Fake Higgs*, although I am not quite as angry (as I am only one-quarter German). Most of what is considered dogma in particle physics can probably be thrown out and started over fresh.

There may be hope for a unifying theory in physics that brings all the loose ends together. "The Theory of Everything" may be a bit ambitious, but we know something better must be out there, because these loose ends don't fit. Someone is going to come up with this. It may be some brilliant, well-published physicist—or, more likely, not. There may be an inherent reason why the present guard has not come up with it to date, as they are fully invested in the present theories that led us down the wrong path at its very start. It is much more likely that some unknown patent clerk comes up with it.

As arrogant as it sounds, this aetherlike view of the universe may be the seed to a grand unification. It contains all of the substance of the theory but needs to be expanded, embellished, edited, and reviewed. The theory is remarkably simple.

The smallest specks of the universe can now be recognized and named. These are the units to which all experiments naturally lead. They allow us to much more easily image in our thoughts when it comes to photons and quarks. Admittedly, it is much more difficult to extend our concepts to Mother Earth or our Milky Way, but the photon and the Milky Way are made of exactly the same specks, nothing added. These are all merely different expressions of Planck unit behavior.

Planck unit behavior is nothing more than the simple yes/no, on/off, 1/0 of the information computer we call our universe. We personally experience this information relay as our reality, but just a tree falling in the woods might need a listener in order to make a sound, reality may be our unique interpretation of these yes/no patterns.

For now, this will have to do until we ascend from the depths to look once more upon the stars.

# BIBLIOGRAPHY

Aczel, Amir D. *Finding Zero*. New York: St. Martin Press, 2015.

Alighieri, Dante. *The Divine Comedy I: Inferno*. Translated by Louis Biancolli. New York: Washington Square Press, 1966.

Baggott, Jim. *Higgs: The Invention and Discovery of the "God Particle."* Oxford: Oxford University Press, 2012.

Bahcall, "New Solar Opacities, Abundances, Helioseismology, and Neutrino Fluxes." *The Astrophysical Journal* 621 (March 2005): L85–88.

Bodanis, David. *E = mc2*. New York: The Berkley Publishing Group, 2000.

Bohm, David. *Quantum Theory*. New Jersey: Prentice-Hall, 1951.

Bohm, David. *Wholeness and the Implicate Order*. New York: Routledge & Kegan Paul, 1980.

Brown, Brandon R. *Planck: Driven by Vision, Broken by War*. Oxford: Oxford University Press, 2015.

Carroll, Sean. *The Particle at the End of the Universe*. New York: Dutton, 2012.

Carroll, Sean. *From Eternity to Here: The Quest for the Ultimate Theory of Time*. New York: Dutton, 2010.

Clausius, Rudolf. *The Mechanical Theory of Heat*. Translated by Walter R. Browne. Cambridge: Macmillan and Co., 1879. Accessed March 10, 2016. University of Notre Dame website.

Davies, Paul. "That Mysterious Flow." *Scientific American* (September 1, 2002). p82-88

Dicke, R. H., P. J. Peebles, P. G. Roll, and D. T. Wilkinson. "Cosmic Black-Body Radiation." *Astrophysical Journal* 142 (July 1965): 414–419.

Dirac, Paul. "The Proton." *Nature* 126 (October 1930): 605–606.

Einstein, Albert, and Leopold Infeld. *The Evolution of Physics*. New York: Touchstone, 1938.

Fayer, Michael. *Absolutely Small*. New York: AMACOM, 2010.

Feynman, Richard P. *QED: The Strange Theory of Light and Matter*. Princeton: Princeton University Press, 1985.

Gell-Mann, Murray. *The Quark and the Jaguar*. New York: Henry Holt and Company, 1994.

Gleiser, Marcelo. *A Tear at the Edge of Creation.* New York: Free Press, 2010.

Greene, Brian. *The Elegant Universe.* New York: First Vintage Books, 2000.

Greene, Brian. *The Fabric of the Cosmos.* New York: New Vintage Books, 2004.

Greene, Brian. *The Hidden Reality.* New York: New Vintage Books, 2011.

Guth, Alan H. *The Inflationary Universe.* New York: Basic Books, 1997.

Hawking, Stephen. *A Brief History of Time.* New York: Bantam Books, 1998.

Herman, Rhett. "How Fast Is the Earth Moving?" *Scientific American*, October 26, 1998.

*Holy Bible.* King James Version. Nashville: Thomas Nelson Inc., 1977.

Huygens, Christiaan. *Treatise on Light.* Translated by Silvanus P. Thompson. Chicago: University of Chicago Press.1690 Accessed March 10, 2016. Project Gutenberg eBook.

Johnston, Alva. "Scientist and Mob Idol—II." *The New Yorker*, December 9, 1933.

Kamenov, Kamen George. *Space, Time, and Matter, and the Falsity of Einstein's Theory of Relativity.* San Bernardino: Vantage Press, 2008.

Kelly, Mervin J. "Clinton Joseph Davisson 1881–1958." In *Biographical Memoirs Vol XXXVI.* New York: Columbia University Press (for the National Academy of Sciences), 1962.

Kirchhoff, Gustav. "On the Relation between the Radiating and Absorbing Powers of different Bodies for Light and Heat." Translated by F. Guthrie in *The London, Edinburgh and Dublin Philosophical Magazine and Journal of Science* from *Annalen der Physik*, July 1860.

Krzywicki, Andre, and Phillipe Dennery. *Mathematics for Physicists.* New York: Harper and Row, 1967.

Kuhn, Thomas S. *The Structure of Scientific Revolutions.* Chicago: The University of Chicago Press, 1962.

Kumar, Manjit. *Quantum: Einstein, Bohr, and the Great Debate about the Nature of Reality.* New York: W. W. Norton & Company, Inc., 2008.

Maxwell, James Clerk. *The Scientific Papers*, edited by W. D. Niven. New York: Dover Publications, Inc., 1890. Accessed March 10, 2016, via Archive.org.

Nambu, Yoichiro. "Spontaneous Symmetry Breaking in Particle Physics: A Case of Cross Fertilization." Nobel Lecture, Stockholm, Sweden, December 8, 2008.

Newton, Isaac. *Opticks.* Fourth Edition. Produced for electronic publication by Suzanne Lybarger, Steve Harris, Josephine Paolucci, and the Online Distributed Proofreading Team. London, 1730. Accessed March 10, 2016. Project Gutenberg eBook.

Peebles, P. J., and Bharat Ratra. "The Cosmological Constant and Dark Energy." *Reviews of Modern Physics* 75, no. 2: Jul 16, 2002 559–606.

Peirce, Penney. *Frequency the Power of Personal Vibration.* New York: Atria Paperback, 2009.

Pickering, Andrew. *The Mangle in Practice: Science, Society, and Becoming.* Durham: Duke University Press, 2008.

Planck, Max. *The Origin and Development of the Quantum Theory.* Translated by H. D. Clark and L. Silberstein. Oxford: The Clarendon Press, 1922.

Planck, Max. *Scientific Autobiography and Other Papers.* New York: Philosophical Library, 1949. Amazon Kindle electronic book.

Polchinski, Joseph. *String Theory.* Cambridge: Cambridge University Press, 2005.

Rees, Martin. *Just Six Numbers: The Deep Forces That Shape the Universe.* New York: Basic Books, 2000.

Rigden, John S. *Einstein 1905: The Standard of Greatness.* Cambridge: Harvard University Press, 2005.

Rochberg, Francesca. "A Consideration of Babylonian Astronomy within the Historiography of Science." *Studies in the History and Philosophy of Science* 33 (2002): 661–684.

Rosenblum, Bruce, and Fred Kuttner. *Quantum Enigma.* Oxford: Oxford University Press, 2006.

Sample, Ian. *Massive.* New York: Basic Books, 2010.

Smolin, Lee. *The Trouble with Physics.* New York: First Mariner Books, 2006.

Smolin, Lee. *Three Roads to Quantum Gravity.* New York: Basic Books, 2001.

Smolin, Lee. *Time Reborn: From the Crisis in Physics to the Future of the Universe.* New York: First Mariner Books, 2014.

Susskind, Leonard, and James Lindesay. *The Holographic Universe.* Singapore: World Scientific Publishing, 2005.

Talbot, Michael. *The Holographic Universe.* New York: Harper Perennial, 2011.

Thomas, Andrew. *Hidden in Plain Sight.* Swansea: Aggrieved Chipmunk Publications, 2012.

Unzicker, Alexander. *The Higgs Fake: How Particle Physics Fooled the Nobel Committee.* CreateSpace Independent Publishing Platform, 2013.

Updike, John. "The Valiant Swabian." *The New Yorker*, April 2, 2007.

Weinberg, Steven. *The First Three Minutes: A Modern View of the Origin of the Universe.* New York: Basic Books, 1993.

Woit, Peter. *Not Even Wrong.* New York: Basic Books, 2007.

Zwiebach, Barton. *A First Course in String Theory.* Cambridge: Cambridge University Press, 2009.

# NOTES

1. From *Enuma Elish*, the Babylonian creation story.

2. See Plato's *Timaeus*.

3. Kuhn, *The Structure of Scientific Revolutions* (Chicago: The University of Chicago Press, 1962), page 10.

4. Kuhn, *Structure*, 5.

5. Johnston, *The New Yorker*, "Scientist and Mob Idol—II," *New Yorker*, 1933.

6. Kuhn, *Structure*, 24.

7. Kuhn, *Structure*, 62.

8. Among these correspondences are Einstein's letters to Schrodinger.

9. Kuhn, *Structure*, 5.

10. From Huygens's *Treatise on Light*.

11. Maxwell, *The Scientific Papers* (New York: Dover, 1890), 500.

12. Kuhn, *Structure*, 7.

13. Brown, *Planck: Driven by Vision, Broken by War* (Oxford: Oxford University Press, 2015), 40–41.

14. Kirchhoff, *Annalen der Physik*, "On the Relation..."

15. Brown, *Driven*, 6.

16. Brown, *Driven*, 26.

17. Brown, *Driven*, 7.

18. Planck, *Scientific Autobiography and Other Papers* (New York: Philosophical Library, 1949), 19–20.

19. Kumar, *Quantum: Einstein, Bohr, and the Great Debate about the Nature of Reality* (New York: W. W. Norton & Company, Inc., 2008), 10.

20. Planck quoted in Brown, *Driven*, 40.

21. Brown, *Driven*, 39.

22. Planck Nobel lecture.

23. From Planck's *New York Times* obituary.

24. Brown, *Driven*, 130.

25. Planck *NYT* obituary.

26. Brown, *Driven*, 119.

27. Planck, *Scientific Autobiography*, eBook location 495.

28. Ibid., eBook location 440.

29. Brown, *Driven*, 122.

30. Ibid., 128.

31. Planck, *Scientific Autobiography*, 33–34.

32. Brown, *Driven*, xii.

33. Ibid., 131.

34. Ibid., 40.

35. Planck quoted in Brown, *Driven*, 127–28.

36. Brown, *Driven*, 128.

37. Rigden, *Einstein 1905: The Standard of Greatness* (Cambridge: Harvard University Press, 2005), 82.

38. Ibid., 82–83.

39. Ibid., 83.

40. Ibid., 92.

41. Ibid., 87.

42. Einstein quoted in Rigden, *Einstein 1905*, 103.

43. Einstein quoted in Rigden, *Einstein 1905*, 19.

44. Rigden, *Einstein 1905*, 19.

45. Ibid., 131–133.

46. Huygens, *Treatise on Light* (Chicago: University of Chicago Press), 16–17.

47. Ibid., 20.

48. Ibid., 19.

49. Greene, *The Elegant Universe* (New York: First Vintage Books, 2000), 192.

50. Planck Mission Team 2013 results.

51. Carrol, *From Eternity to Here: The Quest for the Ultimate Theory of Time* (New York: Dutton, 2010), 58.

52. Brown, *Driven*, 132.

53. Peebles, "The Cosmologic Constant and Dark Energy," *Reviews of Modern Physics* 75, no. 2: 559–606.

54. Dicke et al., "Cosmic Black Body Radiation," *Astrophysical Journal* 142 (July 1965): 414–419.

55. Greene, *Elegant*, 145.

56. Ibid., 142.

57. Greene, *The Fabric of the Cosmos* (New York: New Vintage Books, 2004), 211.

58. Rees, *Just Six Numbers: The Deep Forces That Shape the Universe* (New York: Basic Books, 2000).

59. See Lee Smolin, *Three Roads to Quantum Gravity.*

60. Fully explained in Rhett Herman's "How Fast Is Earth Moving?" in *Scientific American.*

61. See the work of John Wheeler (1955).

62. Explained nicely in the work of Brian Greene.

63. See Weinberg, *First Three Minutes*, 86.

64. Gleiser, *A Tear at the Edge of Creation*, 115.

65. Kelly, "Clinton Joseph Davisson 1881–1958" in *Biographical Memoirs*.

66. Greene, *Elegant*, 105.

67. See Rosenblum and Kuttner's book *Quantum Enigma*.

68. Gell-Mann, *The Quark and the Jaguar* (New York: Henry Holt and Company, 1994), 182.

69. Dirac in *Nature*, 1930.

70. Bahcall, "New Solar Opacities, Abundances, Helioseismology, and Neutrino Fluxes," *The Astrophysical Journal* 621 (March 2005): L85–88.

71. Bohm, *Wholeness and the Implicate Order* (New York: Routledge & Kegan Paul, 1980), 173.

Made in the USA
Middletown, DE
02 June 2016